PROJECT SEA TREE

HARVESTING WIND POWER USING SAILPLANES TETHERED TO A MOVING ROOT

By Dennis Wakefield Stevens

ISBN-10: 146366611X
ISBN-13: 978-1463666118

Table of Contents

APPENDIXES

1. What Are We Going To Do?

Well…there is a lot of fuss about green energy. Why not look at a new and economically attractive way to harvest energy from the wind? There is more than enough energy there to power the entire earth at today's rate of consumption. It is one of the "renewable" energy sources that have attracted attention lately. So, a new, money-making idea should be well received.

How about using kites? Kites have been used for propulsion for hundreds of years. It is said that, in the fifteenth century, Leonardo da Vinci used a kite to traverse a river. Benjamin Franklin once wrote (see http://members.bellatlantic.net/~vze26db3/Miscellaneous/Ben_Franklin.htm) "…I found that by lying on my back and holding the stick in my hands, I was drawn along the surface of the water in a very agreeable manner. Having then engaged another boy to carry my clothes round the pond, to a place which I pointed out to him on the other side, I began to cross the pond with my kite, which carried me quite over without the least fatigue and with the greatest pleasure imaginable." Since then, traction kites have been developed for sport and even to propel cargo ships (see http://www.dcss.org/speedsl/KiteTugs.html and http://www.skysails.info/english/). If they can propel people and machines, why not use traction kites to produce energy?

The purpose of this study is to present a concept for harvesting energy from the wind using special traction kites, to analyze their performance, and to establish the feasibility of the concept. These traction kites are special, because they utilize high performance air vehicles (high lift/drag ratio, high strength, light weight) and high performance surface vehicles (low retardation forces, high resistance to lateral skidding or drift). It is not the purpose of this study to completely design an all-up finished system. Several design features will be suggested that might enhance the performance, and enhance the feasibility of the concept, but the work and dedication required to reduce this to practice is left to all of the Bright Young Light Bulbs (BYLBs) out there. They will be handsomely rewarded for their efforts.

Several people are working on harvesting wind energy using kites. But, they are all looking at kites tethered to the surface of the earth at a fixed point. This point, the tether root, has not been allowed to move, as it does in traction kites. The faster, the better. The faster the root moves, the more wind energy is encountered and is available for harvesting. Kites that move round and round in a somewhat circular path or in a somewhat convoluted circuit, do not count, because there is some point within their system that does not move across the surface of the earth. Furthermore, aerodynamic elements of such machines traversing the windward portion of the circuit cause turbulence. Then, elements traversing the lee portion of the circuit encounter the air contaminated by the machine itself, thereby reducing the wind energy available for harvesting. The horizontal-axis wind turbine (HAWT) does not have that problem (and is a fixed-root system), because the rotor elements sweep out a vertical plane. The rotor blades are never upwind or downwind relative to one another. However, HAWTs are typically distributed in an array. Those on the lee side of the array are exposed to "dirty" air. The moving-root systems discussed here are exposed only to "clean" air. We need to understand the potential of moving-root designs to produce power.

• One of the things we will do is to perform an equilibrium vector analysis (EVA) for the forces on an air vehicle and on a surface vessel (i.e. the moving root) to which it is tethered. This analysis will also be put into a computer program in the form of a spreadsheet. This will permit candidate designs to be evaluated numerically. Don't be concerned if EVA turns you off. It will be in an appendix, and its significance will be summarized.

Well…if we are going to allow the root to move, perhaps we might allow it to move over the ocean. The ocean air is said to be "clean". There are no hills, trees or buildings to disrupt the free flow of the wind, and wind velocities are higher, less turbulent and steadier than over land. Furthermore, the oceans are free. There are no land owners out there. No rent, no lease fees, no purchase price, no government telling you what you must or must not do. That's quite an advantage over land-based or shallow-water-based wind turbines.

But, there are problems. If you are far out to sea, you are far from populated centers and industry that can use the power. It would be absurd to use wind turbines on a sea vessel to generate electrical power in abundance, and have no economical and viable method to transport the energy to where it can be used. Well…when you are surrounded by lemons, make lemonade. When you are surrounded by water with an abundant supply of electrical power, and with many industries trying to dispose of waste carbon dioxide, combine these ingredients, and chemically store the power in the form of a hydrocarbon. So, we will take energy from the environment (convert wind energy to electrical energy) and use it and a catalyst to combine readily available water and carbon dioxide into a hydrocarbon for transport ashore. Much as a tree undergoing photosynthesis, takes energy from the environment (sunlight) and uses it along with a catalyst (chlorophyll) to combine readily available water and carbon dioxide into hydrocarbon tissue. Then, this concept is like a tree at sea, and this study is labeled Project Sea Tree.

• One of the things we will do is to outline a process for catalytically synthesizing hydrocarbon for transport ashore.

But, there are other problems doing this at sea. Moving the surface vessel at high speed at sea will result in high drag forces, and motion against these forces represents wasted power. We are in this game to produce power, so, it is an obvious disadvantage to waste power. But, it goes further than that. Hull drag also limits the velocity of the surface vessel, and the faster it travels the more power is available to be harvested. Consequently, reduction of hull drag is extremely important.

• One of the things we will do is to outline an innovative hull design referred to as the Wheel Keel, which is intended to reduce hull drag.

One of the problems for sailing vessels is the moment arm between the center of aerodynamic lift of the sail and the point of lateral support that resists the lateral force pushing the vessel downwind. This causes the vessel to heel over, reducing the effective sail area and the effective keel area. Lift from the sail also then has a downward component resulting in increased hull drag. The use of a kite to propel the sailing vessel, instead of a fixed sail, can reduce this heeling moment, by bringing the tether root down, close to the keel's center of hydrodynamic pressure. But, that is below the waterline, so the moment is not easily eliminated.

• One of the things we will do is to outline an innovative design, referred to as the Air Keel, which counteracts the heeling moment.

Well…if we are going to let the root move over the ocean, can it not also move over land? We will look at designs that are capable of setting new speed records for land sailing. They can also be used for the generation of power, for transportation, and for just plain fun. In order to model, and to design such systems, we need to understand the retardation forces on land vehicles.

• One of the things we will do is to model friction and drag in land vehicles. Friction and drag measurements in an existing land vehicle for analysis in the model will also be done.

Now…let's shift our attention to the tethered air vehicle. For millennia, kite fliers have battled the problem of instability. There is great anguish when your kite noses over, and crashes into the ground. Even if the design results in stable flight, very often the stability occurs only over a narrow range of wind speeds. Designers have tried using dihedral (wing spar tilted or bowed upward toward the dorsal side of the air vehicle), bridles, multiple tethers, low aspect ratio (span squared divided by wing area) and tails trailing behind the kite. None of these design features are conducive to the production of power from moving-root kites.

• One of the things we will do is to outline an innovative design, referred to as the Pendulum Nose Rudder (PNR), which uses gravity to provide yaw stability.

Is the design of the air vehicle important to the production of power from moving-root kites? Which design features influence power production? Well…as mentioned above, we will look at the characteristics of existing surface vehicles. Maybe we should start by also looking at existing air vehicles.

• One of the things we will do is to collect specifications for several sailplanes. The sailplane designs will be oriented toward a variety of design goals, such as aerobatics, or high-performance, long-distance gliding. These vehicles will be analyzed for their effectiveness in a moving-root kite system using the EVA model mentioned above.

The above tasks should provide some feeling for the feasibility of this approach, and for the importance of various design features. But, the air and surface vehicles are existing vehicles, and they were not designed for this application.

• One of the things we will do is to modify the designs of the best of the existing air and surface vehicles. This will be done to provide a design baseline (DBL) using an unmanned air vehicle under radio control. The DBL will be characterized using the EVA model mentioned above.

OK…what you see is what you get. But, this concept can be enhanced by improving upon certain design features. The performance of the surface vehicle can be improved by reducing the forces of retardation, and by increasing its resistance to lateral forces which tend to push it downwind or to cause sideways skidding. The performance of the air vehicles can be improved by increasing the load limit of the wing spar. Additional air-vehicle improvements would come from reducing its weight and increasing its lift/drag ratio.

• The final thing we will do is to suggest future design improvements. Resistance of the surface vehicle to lateral forces can be improved by putting it on a rail. The techniques of dynamic soaring (DS-ing) will be discussed which would yield air-vehicle designs capable of withstanding 40g air loads.

2. Why Is This Concept Astounding?

• THE POWER FROM THIS SYSTEM USING 67.4 SQUARE METERS OF WING AREA WOULD EXCEED THAT FROM 26,409 SQUARE METERS (OVER 6½ ACRES) OF ROTOR AREA IN A HORIZONTAL-AXIS WIND TURBINE.

• A LAND-VEHICLE VERSION OF THIS SYSTEM WOULD BE CAPABLE OF GROUND SPEEDS WELL IN EXCESS OF 90 METERS PER SECOND (201 MPH) WITH A WIND SPEED OF ONLY 6 METERS PER SECOND (13.4 MPH), FAR AND AWAY BREAKING LAND-SAILING SPEED RECORDS

(see http://www.cnn.com/2009/SPORT/04/08/land.sailing.record/index.html#cnnSTCText).

• THE THRESHOLD WIND VELOCITY REQUIRED TO OPERATE THIS SYSTEM IS A LEISURELY, UPHILL-WALKING PACE OF LESS THAN 1.2 METERS PER SECOND (ABOUT 2.7 MPH). IT WILL BE HARD TO FIND DOLDRUMS SO CALM AS TO FAIL THAT REQUIREMENT.

• THE OCEAN VERSION OF THIS SYSTEM TAKES ADVANTAGE OF THE HIGHER-VELOCITY AND STEADIER WIND AT SEA USING ACREAGE WHICH IS VIRTUALLY UNLIMITED IN SCOPE AND COSTS NOTHING.

• THE GENERATED POWER AT SEA IS USED TO SYNTHESIZE HYDROCARBON IN A PROCESS THAT NOT ONLY AVOIDS THE EMISSION OF CARBON DIOXIDE AS A BYPRODUCT, BUT INSTEAD, IT CONSUMES CARBON DIOXIDE AS A FEEDSTOCK.

3. The Moving-Root Kite (MRK) System

This study concentrates on an air vehicle (or vehicles in a train) tethered to a surface vehicle such as a sea vessel or a land vehicle. The air vehicles have moving, control surfaces including ailerons, elevators and rudders. These surfaces are under radio control, and the air vehicles are unmanned. The surface vehicle is free to move along the surface of the earth, and can contain an engine for propulsion. The engine would be used when there is insufficient wind for sailing and also to attain a starting velocity required to obtain wind propulsion from the tethered aircraft. The aircraft are free to move in three dimensions, although, no provision is made in this study for changing the length of the tether. The system described in this paragraph is referred to as the Moving-Root Kite (MRK). It is intended to capture the energy of the wind for the production of usable domestic and industrial power and/or for achieving high velocities desirable for recreational and transportation applications.

4. the Equilibrium Vector Analysis (EVA) Of The MRK

The vectors involved in the MRK include velocity vectors for the movement of the surface and air vehicles, and the movement of the air over the surface and air vehicles, and they include force vectors for weight and aerodynamic lift and drag on the air vehicles, and for retardation forces tending to slow the surface vehicle. There is an additional force vector for the reaction by the tether to the above forces, causing tether tension. Time changes in velocity result in additional forces of acceleration. These forces can have a significant influence on the performance of the MRK, especially when the air vehicle is caused to accelerate or decelerate relative to the tether root. The technique of using these effects has been referred to as "sheeting". There is no reason why the benefits of sheeting cannot be used with the MRK. Since friction and drag in the surface vehicle is a dominant factor, maybe the air speed of the air vehicle can be enhanced by sheeting while maintaining a reduced ground-vehicle speed (and reduced retardation force). With the increase in air speed, comes a concomitant increase in air loads. But, in addition, the cyclic acceleration of the air vehicle carries with it acceleration forces, which may bring the total load close to the load limit for the air vehicle. The MRK system can usually attain adequate air speed without sheeting. However, sheeting may still be desirable to provide stability of the air vehicle. Here though, the analysis of acceleration forces is left for future work.

When the velocity vectors do not change with time, then the sum of all forces on the system is zero, and the condition of mechanical equilibrium is attained. Equilibrium also requires that the masses of the surface and air vehicles not change with time and that the density of the air passing over the vehicles not change with time. The above vectors were analyzed for equilibrium conditions. This analysis is given in Appendix A. The results of the analysis provide the three components of tether tension in the Cartesian coordinate system. These are the component parallel to the heading of the surface vehicle, T_p, providing forward propulsion, the horizontal component perpendicular to the heading of the surface vehicle, T_t, tending to skid the surface vehicle sideways, and the vertical component, T_z, tending to lift the surface vehicle off the surface of the earth. The gross power delivered by the MRK is obtained by multiplying T_p by the velocity of the surface vehicle. The net power is obtained by subtracting the product of the surface-vehicle retardation force with the velocity of the surface vehicle. The analysis was programmed into a spreadsheet. This provides the capability of numerically evaluating the performance of the MRK, and to evaluate the effects of the parameters of the MRK design, the parameters of the analysis (heading relative to the wind, angle of attack and roll angle of the air vehicles, etc.) and wind speed. This EVA spreadsheet is available (for now) at http://mysite.verizon.net/resyz7ib/ .

5. Existing Sailplanes As Candidate MRK Air Vehicles

Implementation of the vector analysis requires that the input parameters be established. We will start with the air vehicle. It will become apparent that the MRK will operate at high velocities. So, soft, flexible kites made from cloth or paper are not appropriate. A rigid airframe and wing are required. It will also become apparent that minimization of drag (both in the air vehicle and the surface vehicle) is very important. Therefore, candidate air vehicles should have a high lift/drag ratio (L/D). This points toward sailplanes. Even though sailplanes are usually piloted, and the MRK is unmanned, the initial goal is to demonstrate the potential of the MRK using existing vehicles. To do that, parameters for several sailplanes were gathered from the internet. The data is summarized below in Table 1. The parameter $C_{D,0}$ is the coefficient of drag when the coefficient of lift is set to zero by adjusting the angle of attack. Very often this parameter, required in the EVA, is not specified, but the maximum L/D is specified. In such cases, the analysis of Appendix B is used to calculate $C_{D,0}$ in terms of the aspect ratio and maximum L/D.

Table 1. SAILPLANE PARAMETERS

NAME	MODEL	MASS (KG)	WING SPAN (M)	WING AREA (M²)	ASPECT RATIO	MAX L/D	$C_{D.0}$	LOAD FACTOR	SPEED LIMIT (M/S)	LOAD LIMIT (N)	LOAD LIMIT/ WING AREA (N/M²)
PILOTED SAILPLANES											
Glasflugel	Hornet	220.00	15.00	9.80	22.96	38.00	0.01192	5.3		11438	1167
MDM-1 FOX		345.00	14.00	12.30	15.93	28.00	0.01524	9.0	78	30460	2476
Sparrowhawk		70.45	11.00	6.50	18.62	37.00	0.01020	5.5	41	3788	582
Swift	S-1	280.00	12.70	11.80	13.67	30.00	0.01139	10.0	80	27468	2328
Schleicher	ASW-20	255.00	15.00	10.49	21.45	43.00	0.00870	5.3	74	13258	1264
Schleicher	ASW-19	250.00	15.00	10.96	20.53	38.50	0.01039	5.3	68	12998	1186
Schleicher	1 – 23	162.70	13.41	13.83	13.00	25.10	0.01548	8.3	55	13311	962
SZD	Puchacz	365.00	16.67	18.16	15.30	30.00	0.01275	5.3	60	18977	1045
SZD-51-1	JUNIOR	225.00	15.00	12.51	17.99	35.00	0.01101	5.3	61	11698	935
SZD-38	Jantar-1	267.00	15.00	10.66	21.11	39.00	0.01041	5.3	79	13882	1302
Rolladen-Schneider	LS4a	238.00	15.00	10.20	22.06	41.00	0.00984	5.3	75	12374	1213
DG300	Elan	245.00	15.00	10.27	21.91	41.00	0.00977	5.3	75	12738	1240
SZD	Jantar-2	343.00	20.50	14.25	29.49	47.00	0.01001	5.3	69	17834	1251
Schempp-Hirth	Nimbus 3	485.00	24.60	16.85	35.91	57.00	0.00829	5.4	76	25692	1525
Schempp-Hirth	Nimbus 4	595.00	26.50	17.86	39.32	60.00	0.00819	5.3	79	30936	1732
Schneider	ES-65 Platypus	400.00	17.70	15.80	19.83	38.00	0.01030	5.3	72	20797	1316
BLANIK	LAK-17A	215.00	18.00	9.80	33.00	47.00	0.01120	5.3	76	11178	1141
RC DS SAILPLANES											
100 Kinetic		5.62	2.54	0.47	13.70			>40	>200	>2205	>4682
Thundermaster	200	27.00	5.08	1.49	17.36				>200		
Thundertaker		6.81	3.20	0.74	13.78			>40	>200	>2672	>3597

The aerobatic sailplanes have load factors ranging up to 10 (i.e. 10 g), but they have aspect ratios typically below 20, leading to L/D ratios inadequate for high-performance, long-distance flight. Those with high aspect ratios have L/D ratios ranging up to 60, and are capable of long-distance flight. However, they have load factors limited to a little over 5. As we shall see, both L/D and load factor are parameters important to the performance of MRK systems. A few unmanned (RC) sailplanes are included. These have wing spans in the range of 2.5 to 5 meters, and are flown using dynamic soaring (DS-ing) techniques, leading to wind-powered flight in excess of 200 meters per second (447 MPH) and air loads in excess of 40 g (see Appendix C.).

6. Existing Candidate Ground Vehicles

Next, we look at two street vehicles which can be characterized and used as existing ground vehicles in the MRK system. The ground vehicle is considered before the sea vessel as a matter of convenience. Even though these vehicles were not designed for this application, the use of existing vehicles will establish a starting idea of feasibility, and should provide some understanding of the parameters important to this concept.

The three components of tether tension applied to the ground vehicle, as mentioned above, are T_p, T_t and T_z. Each of these require ground-vehicle characteristics conducive to MRK performance. The propulsive force, T_p, should encounter minimal forces of retardation, such as friction and aerodynamic drag. The horizontal force transverse to the heading, T_t, tends to produce skidding and sliding, which wastes energy and diverts the heading of the vehicle. So, it should be resisted. The vertical force, T_z, tends to reduce resistance to skidding and sliding, and should also be resisted, except during a leaping reach (see Section 11.1 below). These characteristics of resistance are discussed in Appendixes D and E for two vehicles, a 1992 GMC Sierra truck and a 1992 Geo Metro.

The analysis in Appendix D is performed assuming that the retardation forces on a wheeled ground-vehicle consist only of friction-like rolling resistance, independent of ground velocity, and aerodynamic drag proportional to the square of velocity. Viscose force proportional to the first power of velocity is ignored in this analysis. The appropriateness of this assumption will become apparent in Appendix E, which considers observed retardation forces. The analysis of Appendix D indicates that when all forces on the vehicle are removed except its weight and the retardation forces, and when the ground velocity is divided by a constant characteristic of the vehicle and the air density, then the arc tangent of that ratio decays linearly with time. This analysis will be used in the interpretation of the observed decay in the velocity of the two candidate ground vehicles in Appendix E.

The two candidate ground vehicles mentioned above are described in Appendix E, and data on retardation forces on these vehicles are presented and analyzed. The results are summarized in Table 2. The aerodynamic drag is obtained by multiplying the shown drag constant by the square of the component of apparent air speed parallel to the vehicle heading. It is seen that the retardation forces are 2 to 3½ times larger in the GMC Sierra than in the Geo Metro which is a big advantage for the Geo Metro. However, the GMC Sierra is about 2½ times heavier, and is superior in resisting transverse forces.

Table 2. RETARDATION FORCES ON CANDIDATE GROUND VEHICLES

VEHICLE	ROLLING RESISTANCE (N)	AERODYNAMIC DRAG CONSTANT (N S^2 / M^2)
1992 GMC Sierra	352	1.233
1992 Geo Metro	99.8	0.596

7. Performance Of The MRK Using Existing Air And Ground Vehicles

7.1 Air Vehicles We begin this section by looking at various sailplanes towing a 1992 Geo Metro. The first analysis is for a Nimbus 3 air vehicle. The EVA spreadsheet program requires input of the starting conditions. This includes characteristics of the air vehicle (wing area, wing span, mass, drag coefficient at zero lift, and load factor), characteristics of the ground vehicle (weight, drag constant, and rolling resistance), and environmental conditions (air density of 1.22 Kg/m^3 and wind speed assumed to be 6 m/s). The spreadsheet program requires operator input for the velocity and heading of the ground vehicle. Typically, the heading is 180 degrees (into the wind) until the air vehicle is airborne, and then it is gradually changed to the desired heading as the velocity builds. This requires supplementary propulsion supplied by an engine in the ground vehicle. After the tethered air vehicle provides positive net propulsion (after subtraction of friction and drag), the engine is disengaged. The control of stability of the air vehicle is discussed in Appendix F. The ground velocity is typically incremented by 0.5 M/S for each calculation of flight parameters. The roll and angle of attack of the air vehicle are also programmed by the operator in order to achieve the desired power output and air-vehicle altitude (although angle of attack is typically held constant at 10 degrees to maximize the lift coefficient and to remain below stall). Increasing roll increases power output, because the horizontal component of lift is increased. However, that will also reduce the altitude of the air vehicle, and needs to be controlled by the operator.

The spreadsheet was programmed to calculate time lapse between velocity increments, assuming that the engine provided an acceleration of 1 M/S/S until the propulsion from the air vehicle became positive. Thereafter, acceleration was taken to be the ratio of the net propulsive traction to the mass of the ground vehicle. The time lapse was then taken to be the ratio of the velocity increment to the acceleration, and the distance traveled during the increment was taken to be the product of the time lapse with the average velocity during that increment. The accumulation of the distance increments and the heading provided the path traveled by the ground vehicle. The resultant pathway is shown in Figure 1.

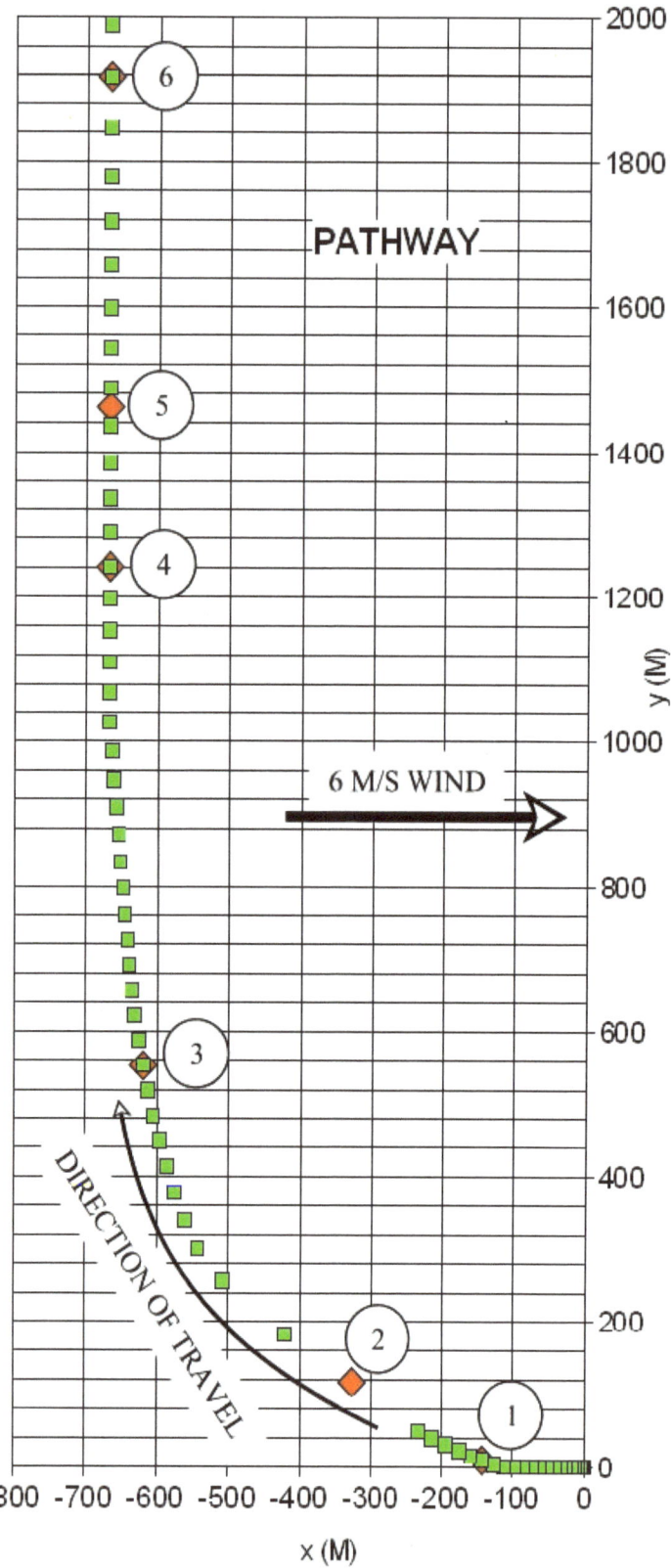

FIGURE 1. PATHWAY TRAVELLED BY A
GEO METRO TOWED BY A NIMBUS 3

The green, square data points in Figure 1 represent the position of the ground vehicle. The positions in Figure 1 are shown for increments in velocity of 1 M/S. The intermediate calculations at increments of 0.5 M/S were not shown for clarity. So, each square data point represents the position when the velocity reaches 1 M/S above that of the previous point. The red, diamond-shaped points are event markers, and the events are described in Table 3. Event 7 is not shown in Figure 1, because it occurred several kilometers down range. Event 8 did not occur in this MRK configuration, but can occur when a train of several air vehicles is used.

Table 3. EVENTS FOR A GEO METRO TOWED BY A NIMBUS 3

Marker	Velocity (M/S)	Heading (Deg)	Roll (Deg)	Angular Altitude (Deg)	Power (W)	Comment
1	17	160	25	1.5	-4,703	Air vehicle lift off. Heading is nearly into the wind.
2	22.5	145	50	2.8	2,133	Power becomes positive. Propulsive traction exceeds friction and drag. Ground-vehicle engine is disengaged.
3	32	100	63	2.8	29,213	Lateral skidding force exceeds weight of ground vehicle.
4	50	90	77	2.5	54,891	About to exceed load limit of air-vehicle spar.
5	54.5	90	79	2.2	57,162	Maximum power.
6	62	90	80	3.2	54,370	Lateral skidding force exceeds 4 times the weight of ground vehicle
7	82.5	90	83	3.1	145	Power almost falls to zero near maximum velocity.
8	Not observed at 99 M/S					Vertical component of tether tension exceeds ground-vehicle weight.

Spreadsheet calculations of forces on the ground vehicle are shown in Figure 2. It shows propulsive traction provided by the tethered air vehicle and the total retardation forces from friction and drag on the ground vehicle. The difference provides the net propulsive force, and that multiplied by the ground speed gives the power developed by the system. The power is shown in Figure 3. The events shown in figures 2 and 3 are the same events shown in Figure 1, and described in Table 3.

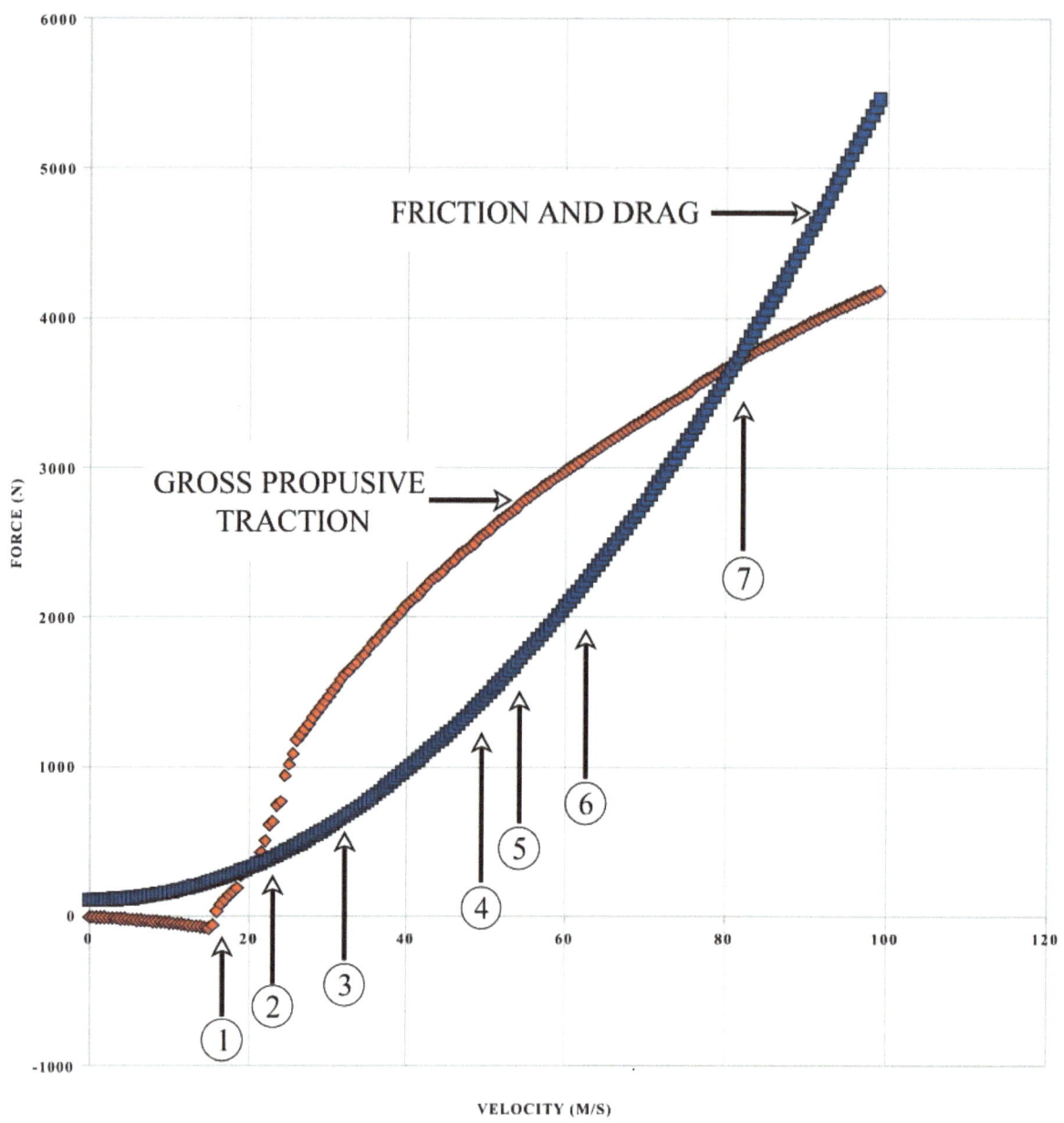

FIGURE 2. PROPULSIVE AND RETARDATION FORCES ON A GEO METRO
TOWED BY A NIMBUS 3

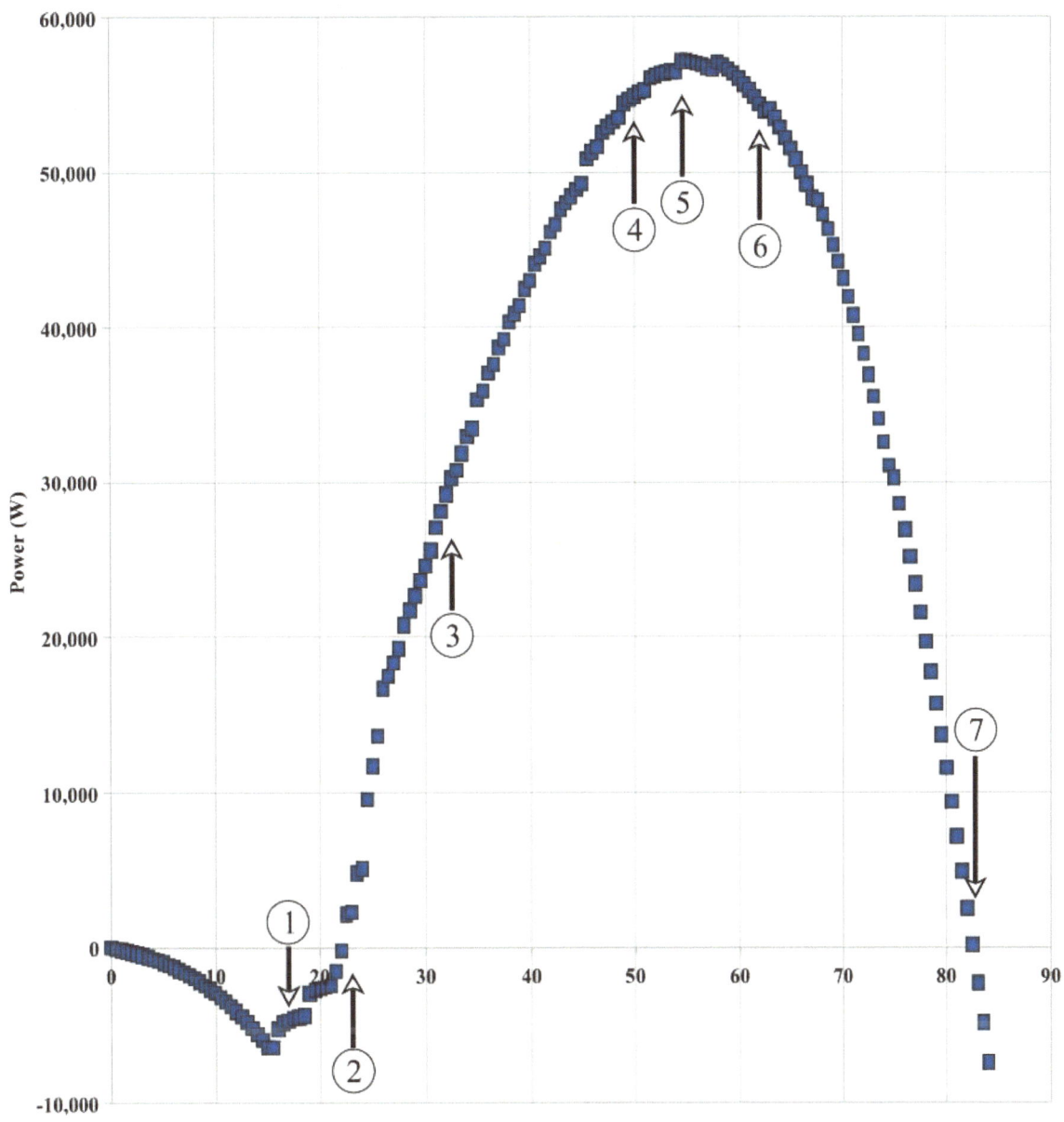

FIGURE 3. NET POWER GENERATED BY A GEO METRO TOWED BY A NIMBUS 3

The lateral skidding force occurs well before the achievement of maximum power. The skidding problem may be delayed by the use of an aft air foil on the ground vehicle. Such an air foil used on race cars has been called an "AeroGrip", and they have allowed race cars to successfully negotiate 4g turns without skidding. So, event 6 (4g lateral force) was included in these figures. An AeroGrip might be expected to cause added drag to the ground vehicle, which was not taken into account here, but it should permit a closer approach to maximum power or power at the load limit than a 1g skid (event 3). It should be noted here that 1g skids or 4g skids refer to skidding due to lateral loads delivered locally, to the root by the tether corresponding to the vehicle weight or four times the vehicle weight, even though the lateral loads are not distributed g-loads.

The structural load limit provides a limit to the generated power and to the velocity of the ground vehicle when the angle of attack of the air vehicle is held at 10 degrees. If the angle of attack is reduced, then lift and air loads are reduced, and so, likewise, is the generated power. But, as long as the reduced net power remains positive, the system can accelerate to higher velocities. Therefore, the EVA spreadsheet was run again using a programmed reduction in angle of attack for the purpose of achieving higher velocities without exceeding the load limit. The result is shown in Figure 4.

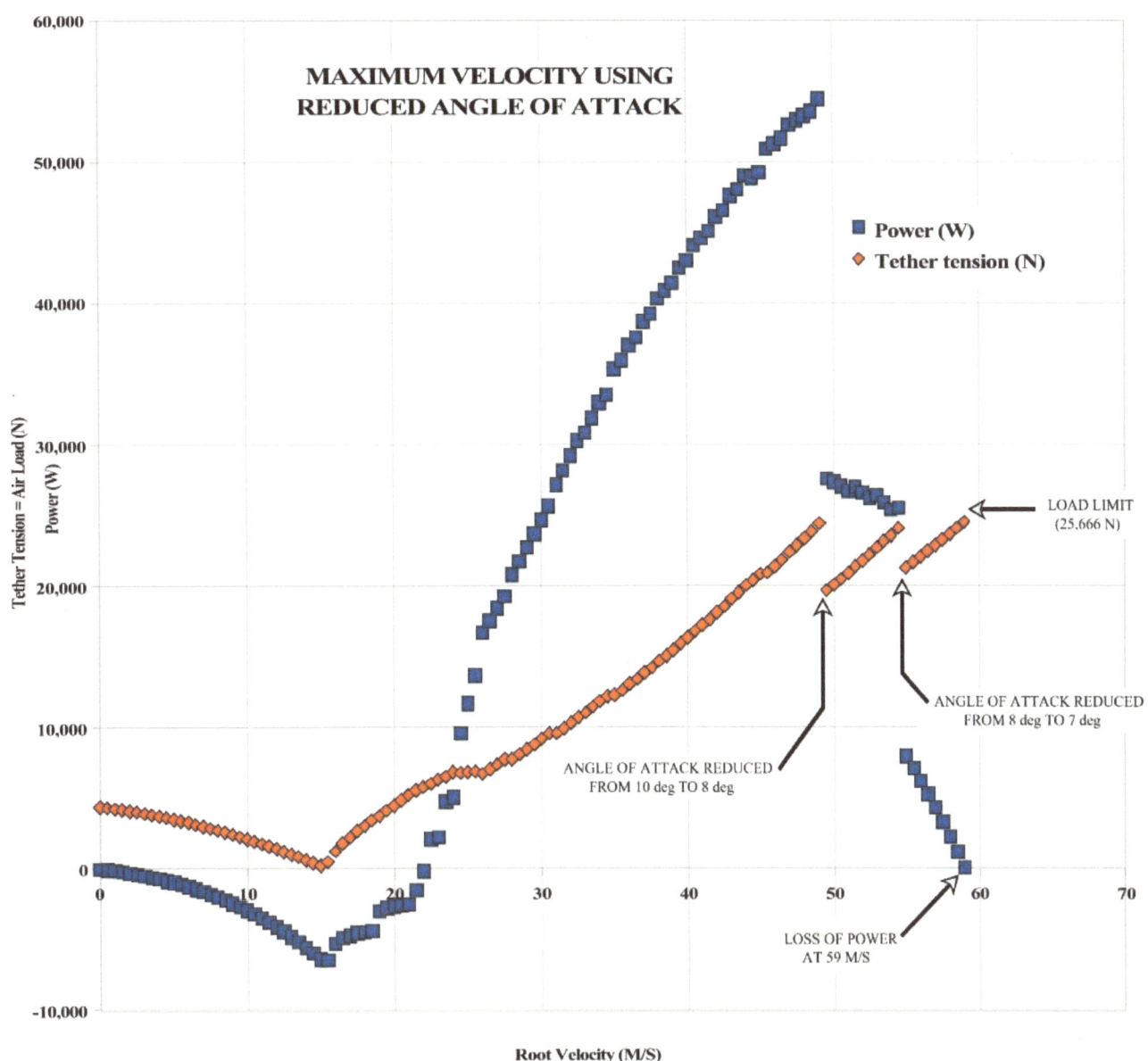

FIGURE 4. USING REDUCED ANGLE OF ATTACK TO ACHIEVE MAXIMUM SPEED OF A GEO METRO TOWED BY A NIMBUS 3

By reducing the angle of attack before the load limit is reached, the velocity can be increased from 50 to 59 M/S without exceeding the load limit and before net power is lost. This with a wind speed of only 6 M/S. This is a modest improvement over the speed with a ten-degree angle of attack, but other configurations result in very significant improvements, as will be seen. It should be noted here that reductions in the angle of attack not only reduce air loads on the wing, but also reduce lateral skidding forces. Now, it is seen that by using a low-drag surface vehicle, and an air vehicle with a high lift/drag ratio, one can kite-sail nearly ten times the speed of the wind. And, that is limited by the load factor of the air vehicle. An improvement there and even higher speeds are attainable. Even better performance will be shown for other MRK configurations. Sailing speed is typically limited to about three times the speed of the wind, but sailing speed on ice can approach five times the speed of the wind at low wind speeds. (See http://iceboat.org/faqiceboat.html , and http://www.idniyra.org/articles/boatspeeds.htm).

A reduction in friction and drag on the surface vehicle would improve performance. For the present though, we will look at increasing the forward propulsion without changing the surface vehicle. This can be accomplished by increasing the wing area of the air vehicle, either by increasing the size of the air vehicle or by adding more vehicles in a train. The Nimbus 3 is presently in consideration for the air vehicle, which has a wing span of 24.6 M (about 80.7 feet). It is already large and cumbersome. Therefore, EVA-spreadsheet calculations were made for the Geo Metro towed by trains of 1, 2, 3 and 4 Nimbus 3 air vehicles, one behind the other. Presumably, they would be connected by three tethers between each adjacent vehicle, two at outboard positions on the wing and one at the tail. If the innermost air vehicle is the master, controlled by radio, then the trailing slave vehicles would be controlled by the tethers. Conceivably, the train would have more stability than an individual vehicle. In that case, control surfaces on the slave vehicles may not be needed, and the entire tail assembly could be eliminated on the slave vehicles. The fuselage would serve only as a tether point at the tail end. These advantages in control and stability with concomitant savings in weight and drag are not taken into account here. The EVA-spreadsheet calculations of forward traction and friction and drag for such trains are plotted versus root velocity in Figure 5, and the performance characteristics are shown in Table 4.

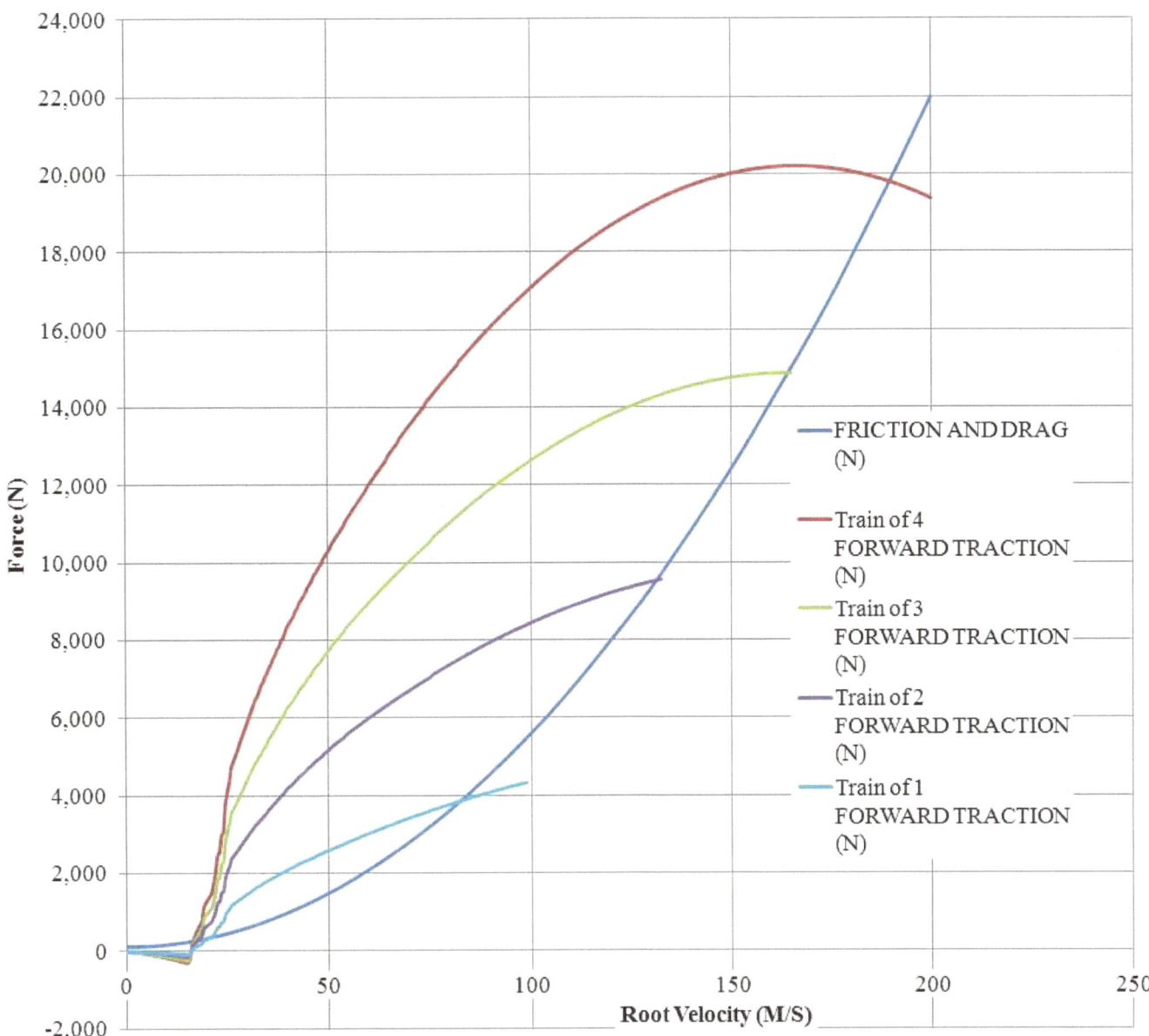

FIGURE 5. Forward Traction, Friction And Drag on a Geo Metro
Towed By Trains of Nimbus 3 Air Vehicles

Table 4. The MRK Performance of a Geo Metro Towed by Trains of Nimbus 3 Air Vehicles

PERFORMANCE	TRAIN OF 1	TRAIN OF 2	TRAIN OF 3	TRAIN OF 4
VELOCITY AT 1g SKID (M/S)	32	25.5	23	22
POWER AT 1g SKID (W)	29,213	38,807	37,241	35,738
VELOCITY AT 4g SKID (M/S)	62	44	36.5	32.5
POWER AT 4g SKID (W)	54,370	148,484	173,715	185,867
VELOCITY AT LOAD LIMIT (M/S)	50	50	50	50
POWER AT LOAD LIMIT (W)	54,891	183,645	312,798	441,950
VELOCITY AT MAXIMUM POWER (M/S)	54.5	90	113	129
MAXIMUM POWER (W)	57,162	323,494	773,637	1,356,542
VELOCITY AT LOSS OF POWER (M/S)	82.5	134	169	193

If skidding occurs at a lateral force equal to the weight of the ground vehicle (1g skid), then it occurs early into the kite-powered flight, and adding air vehicles to the train causes the 1g skid to occur even earlier. However, there is a power benefit in going from one air vehicle to two. Using a 4g AeroGrip provides a greater power benefit in adding vehicles. The size of the train has no effect on the velocity at the load limit for the air vehicles, but a very significant increase in power at the load limit is seen for larger trains. At even higher velocities (if one can solve the skid problem and the load-limit problem) power passes through a maximum. Then, a train of four air vehicles can produce a maximum power of over a megawatt from a total wing area of 67.4 m^2 in a wind speed of 6 m/s. But, if you are more interested in high speed, a single air vehicle will avoid skidding better than a train.

So far, only a wind speed of 6 m/s has been considered. It would be desirable to evaluate how low the true wind need be for the MRK system to operate. Accordingly, the EVA spreadsheet was run for a train of 4 Nimbus 3s towing a Geo Metro with low wind speeds. The resultant calculations of power are shown in Figure 6. The power charts for sequentially lower values of wind speed look like the sun setting on the horizon. So, the wind speed where power disappears below the horizon is referred to as the sunset wind. The chart shows that this MRK configuration has a sunset wind below 1.3 m/s, an extremely small threshold. The other remarkable observation from Figure 6 is that the root velocity at loss of power ranges up to about 30 times the true wind speed. That is vastly greater than observations for land sailing using either kite propulsion or a sail on a fixed mast. Land sailing with a kite has been done with a cloth kite having much poorer performance parameters than those for the MRK systems considered here. The land-sailing speed record was set in 2009 using a fixed-mast, rigid sail that looked like a wing with a high aspect ratio. With a fuselage and tail assembly attached to the wing, it looked like a sail plane tilted up on its wing tip. That was going in the right direction, but it lacked control over wing roll (vertical wing) and was too limited in sail area. It should be noted here that operation above 50 m/s puts the Nimbus 3 above its load limit. That load limit can be delayed out to higher velocities and higher wind speeds using the method described above (see Figure 4.)

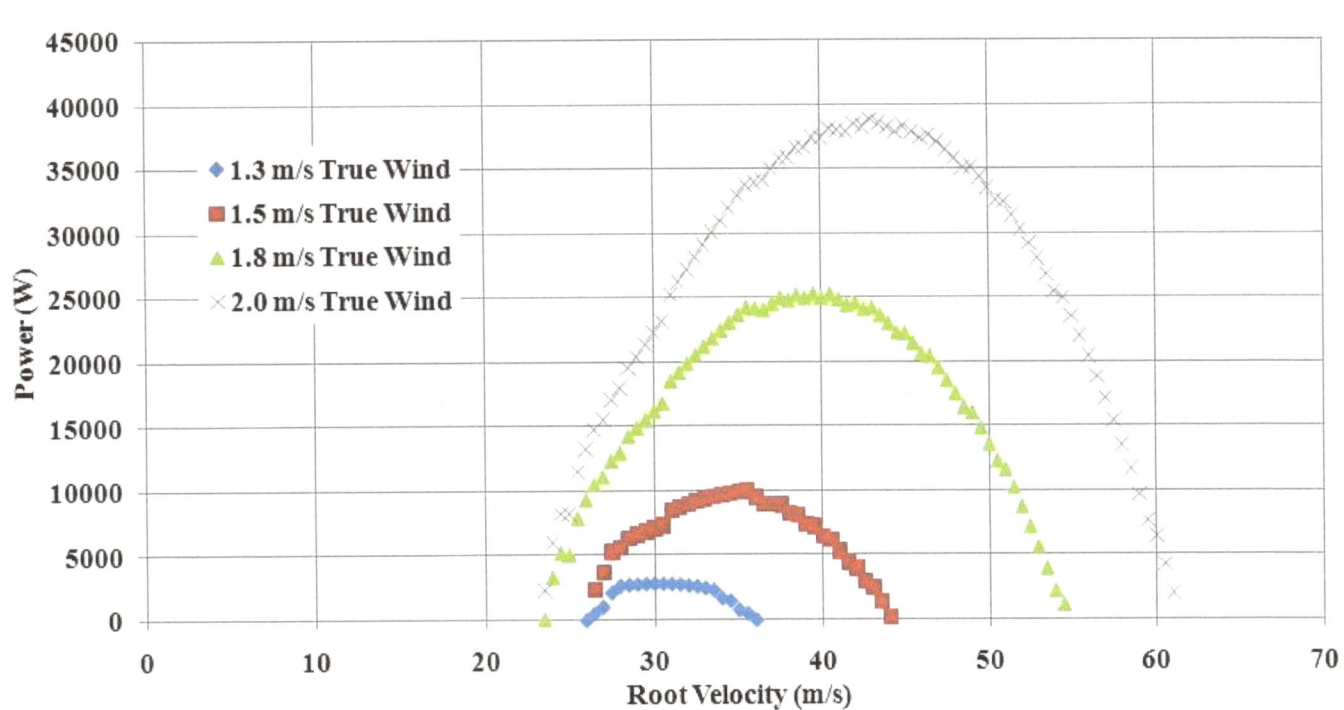

FIGURE 6. Sunset Wind For A Geo Metro Towed By A Nimbus 3

Next, consider several air vehicles, two with high aspect ratios (Nimbus 3 and LAK-17A) and two with high load factors (MDM-1 Fox and Swift S-1). The EVA spreadsheet was run for those vehicles, yielding calculations of maximum power shown in Figure 7 and load-limited power shown in Figure 8. Maximum power was achieved in trains of 1 and 2 vehicles before the load limit was reached with all except the Nimbus 3. So, there are fewer data points in Figure 8 than in Figure 7. Up to 5 LAK-17A vehicles and up to 5 Swift S-1 vehicles were considered. Only 4 were used for the other two vehicle types. Maximum power is highest with the Nimbus 3, indicating that high aspect ratio leading to high lift/drag ratio dominates. Even at its lower load limit, the Nimbus 3 provides load-limited power higher than those of the aerobatic MDM-1 Fox and Swift S-1. The smaller LAK-17A provided the highest load-limited power. It out-performed the Nimbus 3, probably because the LAK-17A with a slightly lower aspect ratio has a wing loading of only 22 Kg/M^2 compared to about 29 Kg/M^2 for the Nimbus 3. A straight line was fit to the load-limited power calculations shown in Figure 8 for the Nimbus 3. Load-limited power is proportional to the first power of wing area achieved by adding vehicles in a train, but maximum power is not. These calculations indicate that when towing a Geo Metro, existing air vehicles with high aspect ratios out-perform those with a high load factor.

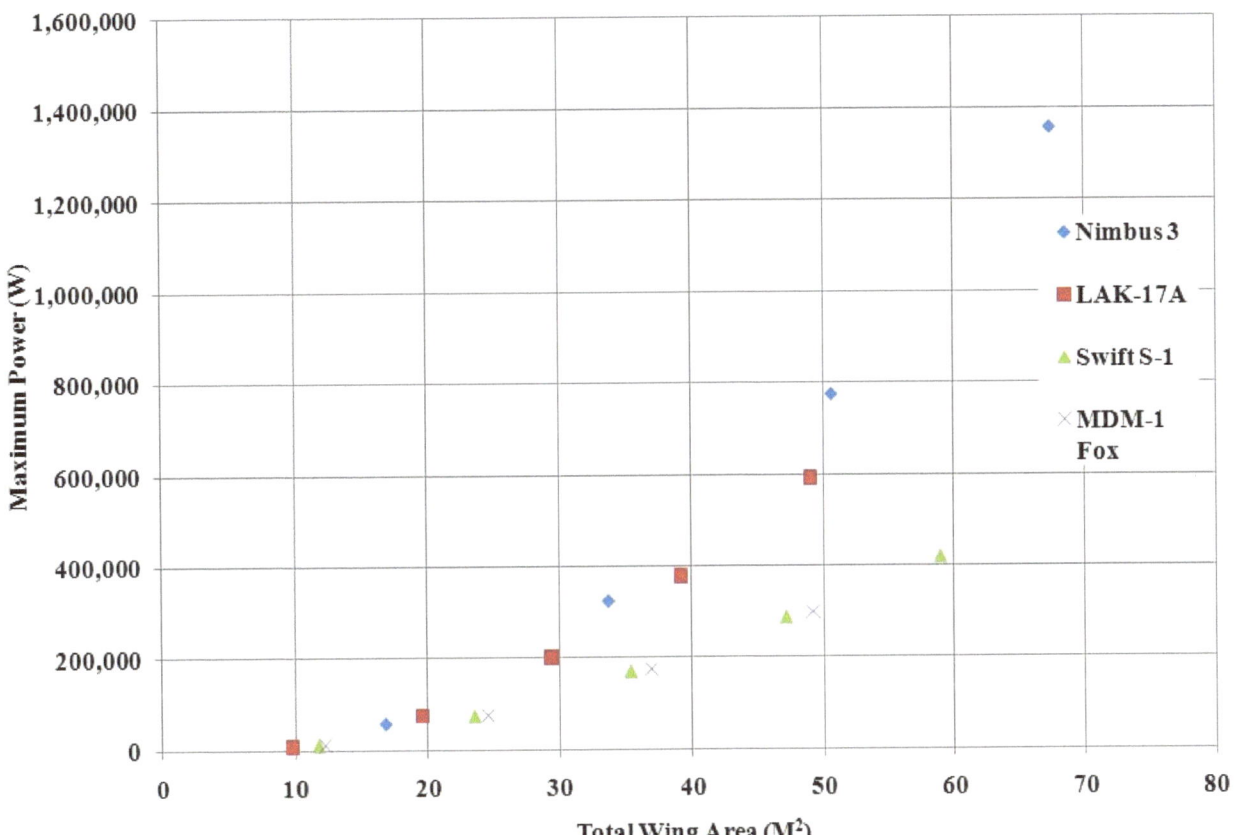

FIGURE 7. Maximum Power Generated By A Geo MetroTowed
By Different Air-Vehicle Trains

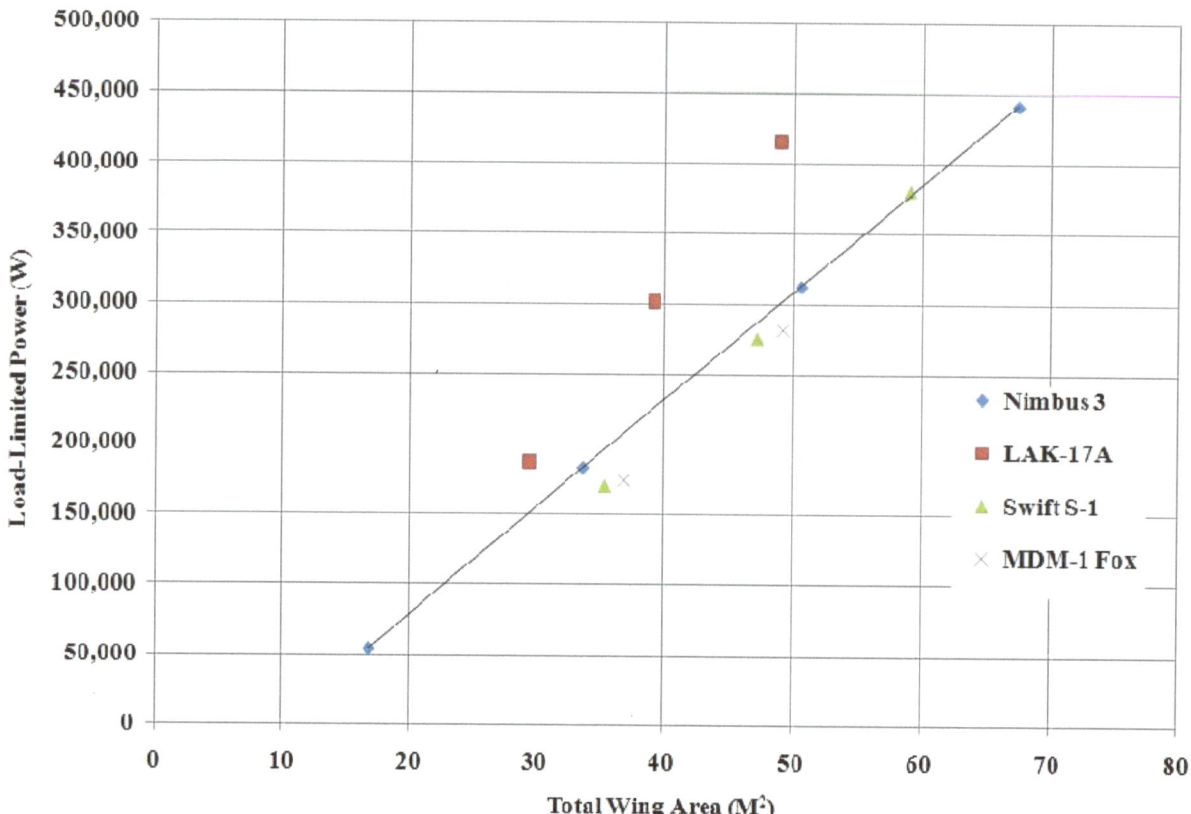

FIGURE 8. Load-Limited Power Generated By A Geo Metro Towed By Different Air-Vehicle Trains

7.2 Ground Vehicles Next, we compare the performance of the ground vehicles, Geo Metro and GMC Sierra when towed by a Nimbus 3. The two ground vehicles differ significantly in that the Geo Metro has lower mass, leading to less friction, and lower frontal area, leading to less aerodynamic drag. Conversely, the GMC Sierra is more massive, and is better at resisting skidding forces. Accordingly, the EVA spreadsheet was run for trains of up to four Nimbus 3s towing the two ground vehicles. The power developed at a 1g skid and at a 4g skid are shown in Figures 9 and 10. The terms "1g skid" and "4g skid" refer to the lateral skidding force equal to the vehicle weight and to 4-times the vehicle weight. These are terms of convenience, but these forces differ from "g" forces in that they are applied locally to the tether attachment point rather than being distributed as would be an inertial force. The GMC Sierra towed by a single Nimbus 3 failed to produce positive power, so there are fewer data points for the GMC Sierra. This failure indicates that the sunset wind for the GMC Sierra towed by a single Nimbus 3 is about 6 M/S.

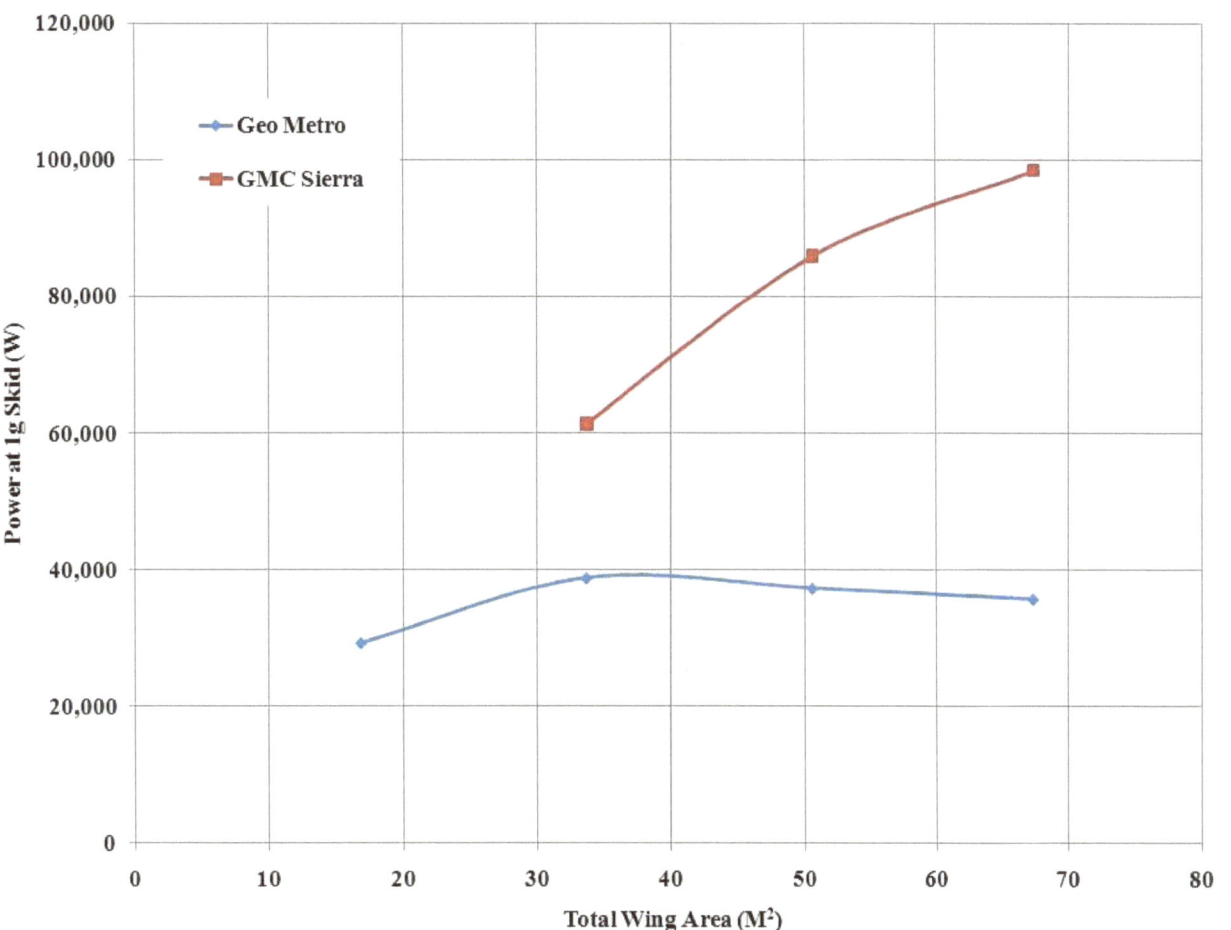

Figure 9. Power From Trains Of Nimbus 3s Towing A Geo Metro
Or A GMC Sierra Into A 1g Skid

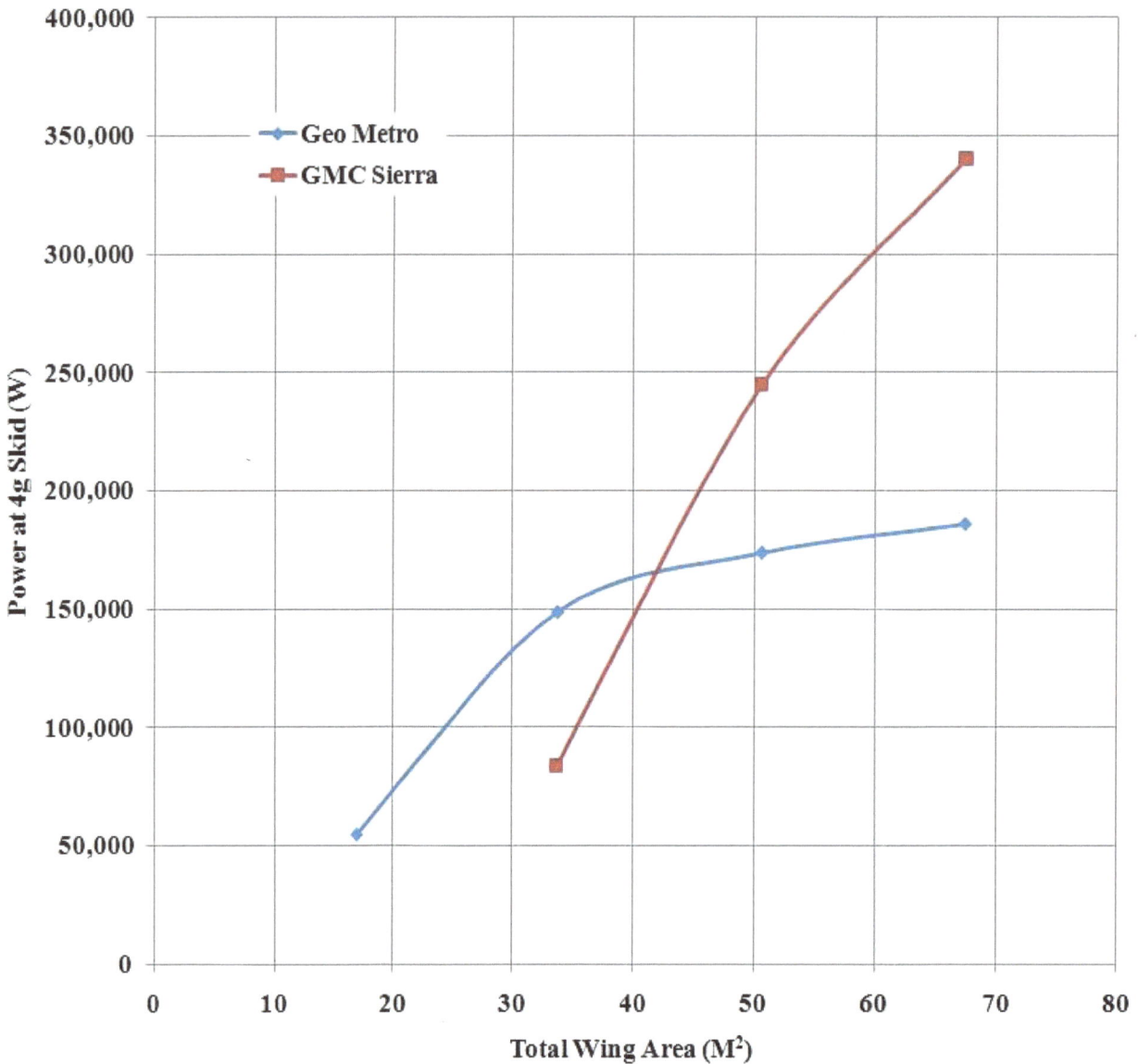

Figure 10. Power From Trains Of Nimbus 3s Towing A Geo Metro Or A GMC Sierra Into A 4g Skid

As indicated in Table 4, the power developed in the Geo Metro at a 1g skid was improved slightly by adding a second air vehicle, but the power fell off for three and four vehicles, because the skid occurred at lower velocities as the wing area was increased. However, the much larger mass of the GMC Sierra delayed both the 1g and the 4g skids to higher velocities, especially for trains of four air vehicles, resulting in large increases in power compared to that of the Geo Metro.

8. Design Modifications

8.1 Air Vehicle Modifications. So far, we have considered only existing vehicles for the MRK system. Those analyses have shown potential of the MRK for the generation of power and the attainment of high velocities. But, as noted above, these vehicles were not designed for this application, and some design modifications should provide improved performance. The air vehicles considered in the above analyses were all manned, and the cockpit is unnecessary for the MRK. The removal of the cockpit will reduce the mass of the vehicle, permitting more roll of the air vehicles (less vertical component of lift for opposing gravity and more horizontal component of lift for propelling the ground vehicle). The lower mass will also permit launch at a slightly lower air speed. The cockpit removal will also reduce the frontal area of the air vehicles which will reduce drag. For the design baseline (DBL), we will consider a modification of the Nimbus 3, since it was a top performer in the above analyses. Figure 11 shows the front profile of the Nimbus 3 compared to a modification taken to be the air-vehicle design baseline. The large reduction in the size of the fuselage results in an overall mass reduction of 10.3% and a reduction of 12% in the entire frontal area of the vehicle. It is assumed that the coefficient of drag at zero lift ($C_{D,0}$) is directly proportional to frontal area, so that corresponds to a 12% reduction in $C_{D,0}$.

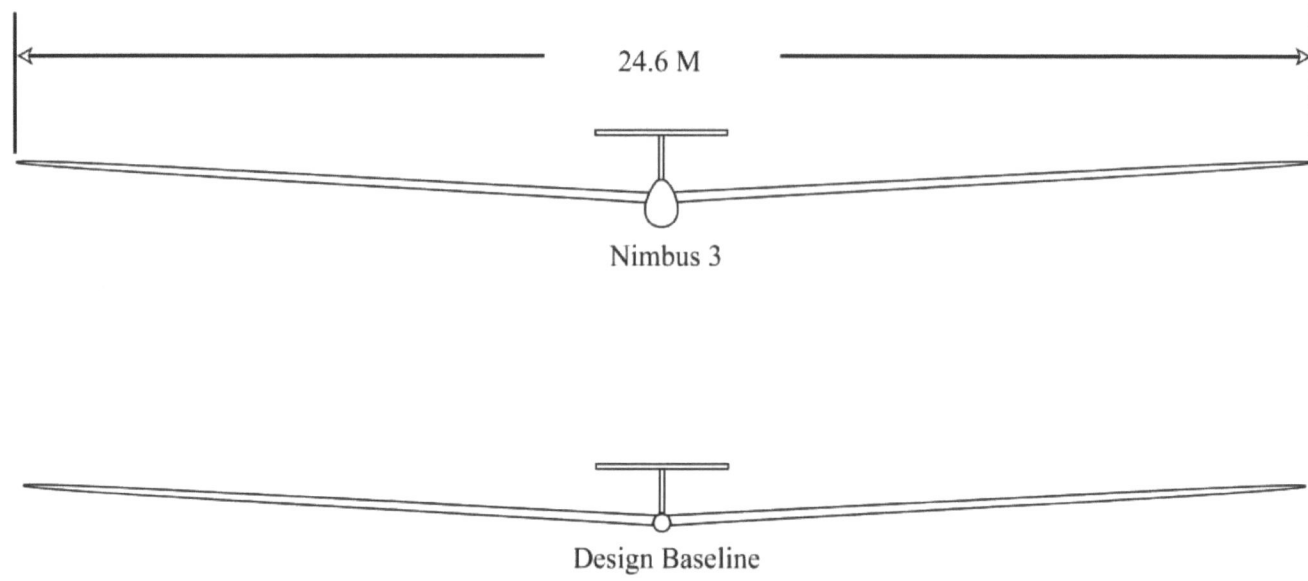

Figure 11. Comparison of Front Profiles of Nimbus 3 and The Design Baseline

An additional design improvement might be obtained by taking into consideration the greatly enhanced load factor, ranging up to at least 40g in DS sailplanes (see Table 1.) However, the high observed load factors are achieved at the expense of lower aspect ratios. If we assume that a trade-off can be made between load factor and aspect ratio, then improved performance at the load limit might be obtained at intermediate values of those parameters. Such a trade-off is shown in Figure 12. The 40g data point represents the DS sailplane 100 Kinetic with an aspect ratio of 13.7, and the 5.4g data point represents the Nimbus 3 with an aspect ratio of 35.91. A linear relationship between these points was assumed, which was used in the EVA spreadsheet. The DBL was modified by reducing the wing span while holding the wing area and the air-vehicle mass constant. This had the effect of changing the aspect ratio and the load factor systematically, in accordance with the relationship shown in Figure 12. The resultant plots of maximum power and power at the load limit versus aspect ratio are shown in Figure 13 for a wind velocity of 6 m/s and in Figure 14 for a wind velocity of 9 m/s. Each spreadsheet calculation was done for a train of four modified DBL air vehicles towing a DBL ground vehicle. The ground vehicle is described in the next section.

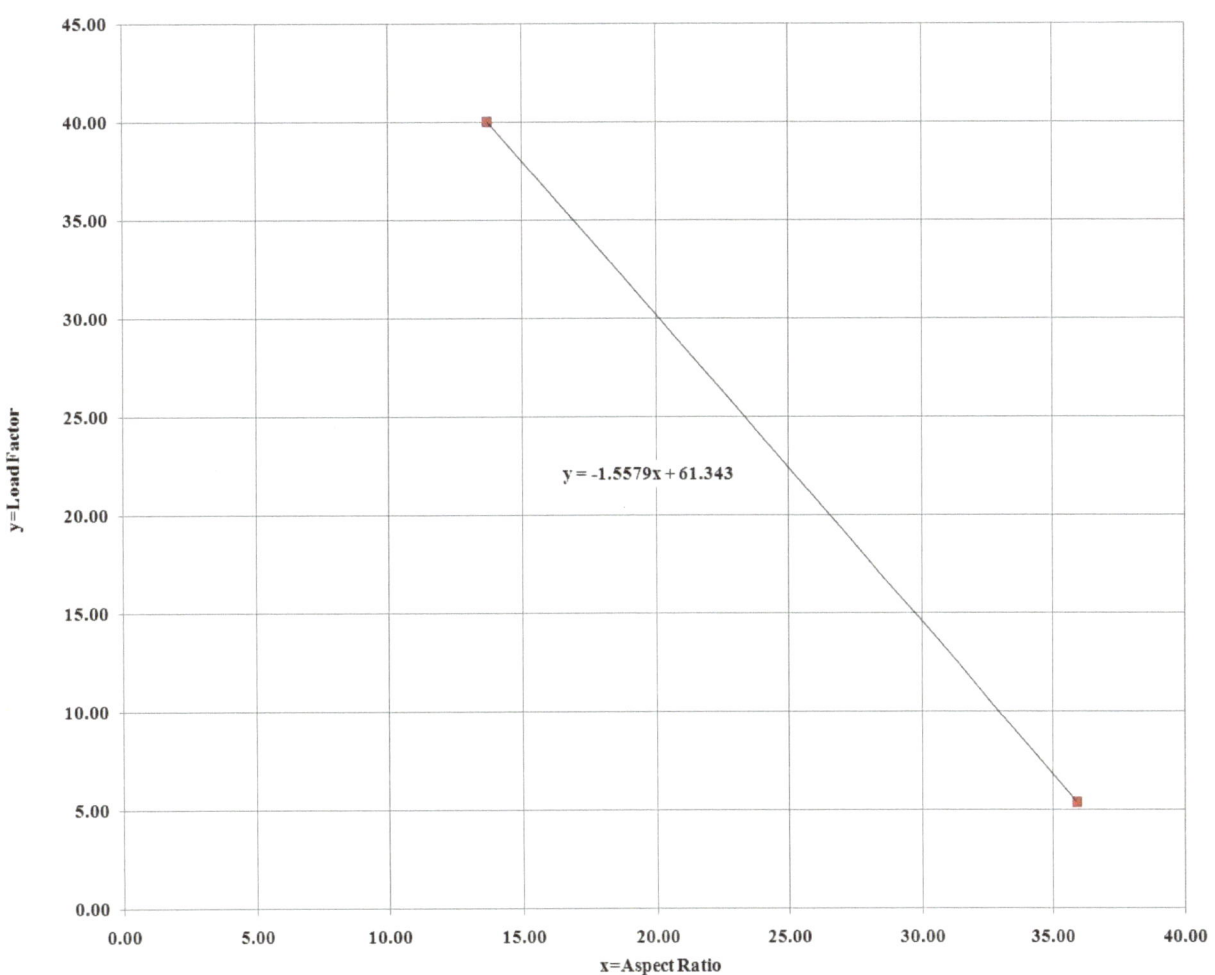

Figure 12. Assumed relationship between Load Factor and Aspect Ratio

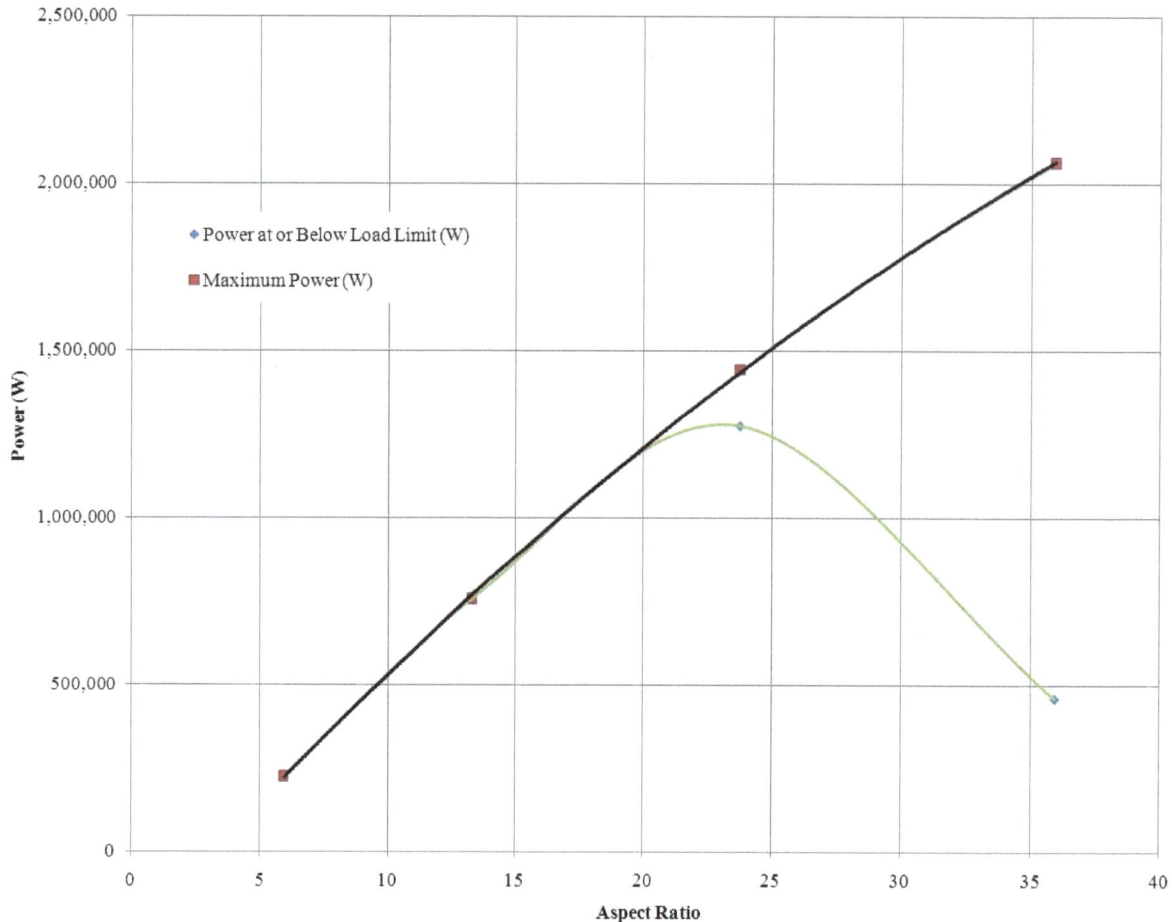

Figure 13. DBL Maximum Power and Power at Load Limit Versus Aspect Ratio in 6 m/s Wind (Load Factor Falls Off With Increasing Aspect Ratio)

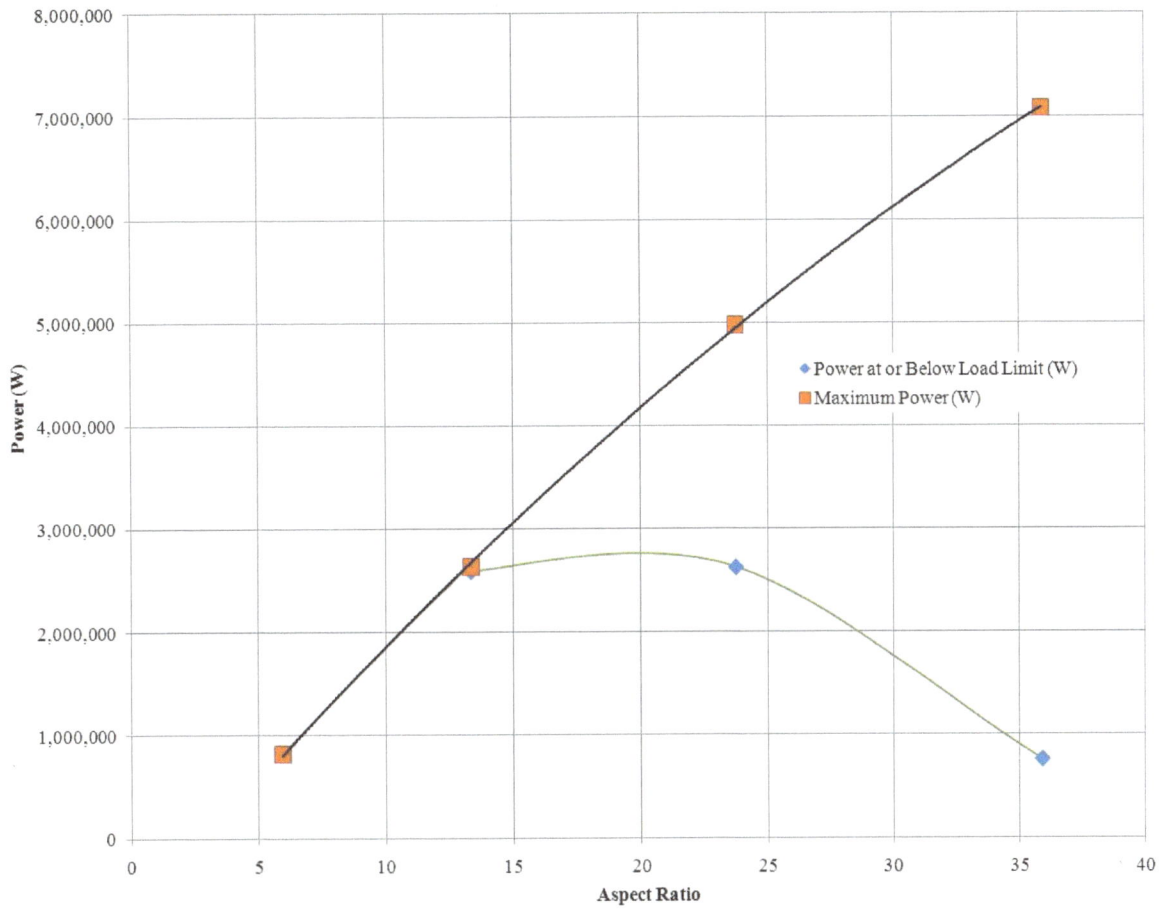

Figure 14. DBL Maximum Power and Power at Load Limit Versus Aspect Ratio in 9 m/s
Wind (Load Factor Falls Off With Increasing Aspect Ratio)

It is seen that maximum power increases with increasing aspect ratio, but power
at the load limit passes through a maximum at an aspect ratio of about 20 to 23.
Therefore, a second design baseline (DBL02) is taken with an aspect ratio of
20. In order to introduce more wing area, the DBL02 wing span is kept at 25 m,
while maintaining the same wing loading of just over 25 Kg/m^2 which results in
a mass of 800 Kg per air vehicle. The characteristics of the DBL air vehicles
are compared with those of the Nimbus 3 in Table 5. Note that while DBL02
almost doubles the wing area of DBL, the load limit is increased by almost an
order of magnitude.

Table 5. Characteristics of DBL and DBL02 Air Vehicles Compared With Nimbus 3

Name	MASS (KG)	WING SPAN (M)	WING AREA (M²)	WING LOADING (KG/M²)	ASPECT RATIO	MAX L/D	$C_{D,0}$	LOAD FACTOR	LOAD LIMIT (EACH AIR VEHICLE) (N)	LOAD LIMIT/ WING AREA (N/M²)
Schempp-Hirth Nimbus 3	485	24.6	16.85	28.78	35.9	57.00	0.008291	5.4	25,692	1,525
DBL Air Vehicle	435	24.6	16.85	25.81	35.9	60.74	0.007300	6.0	25,692	1,525
DBL02 Air Vehicle	800	25.0	31.25	25.60	20.0	45.33	0.007300	30.2	236,690	7,574

8.2 Ground Vehicle Modifications. The calculations shown in Figures 9 and 10 indicate that the GMC Sierra outperforms the Geo Metro when towed by a train of air vehicles. The superior resistance to skidding by the GMC Sierra resulted from its much higher mass. So, a modified GMC Sierra is selected for the DBL ground vehicle. The mass is taken to be unchanged which maintains the resistance to skidding, but the frontal area (2.8 m²) and the coefficient of drag (0.8) are significantly reduced in order to reduce aerodynamic drag. Review of the internet literature indicates that a typical automobile has a drag coefficient of around 0.34 and a product of drag coefficient with frontal area equal to about 6 square feet (0.557 m²). That corresponds to a frontal area of 1.64 m². The selected DBL parameters are compared to those of the GMC Sierra in Table 6. The term ½ $\rho A C_D$ multiplied by the square of the air speed, yields aerodynamic drag. Here, ρ is the air density (1.22 Kg/m³), A is the frontal area and C_D is the coefficient of drag.

Table 6. DBL Ground Vehicle Compared to 1992 GMC Sierra

NAME	MASS (kg)	ROLLING RESISTANCE (N)	FRONTAL AREA (m²)	COEFICIENT OF DRAG, C_D	½$\rho A C_D$ (kg/m)
1992 GMC SIERRA	2588	352	2.8	0.8	1.34
DBL ground vehicle	2588	352	1.64	0.34	0.34

In summary, the DBL consists of four DBL air vehicles Towing the DBL ground vehicle, and the DBL02 consists of four DBL02 air vehicles towing the same DBL ground vehicle.

9. Performance Of The Design Baseline

The EVA spreadsheet was run for the DBL. Calculated power versus root velocity is shown in Figure 15 with event markers defined in Table 7. In comparison with the Nimbus 3 (see Table 3. with the same event markers), the DBL air vehicles with a lower wing loading, are launched at a lower velocity than the Nimbus 3. Because of the lower wing loading and less drag, the root velocity required to achieve positive power is also less than that for the Nimbus 3. In comparison with a train of four Nimbus 3s (see Table 4.), DBL skidding is delayed to a higher velocity, resulting in much higher power at the 1g and 4g skids. This delay to higher velocity at skid occurs because of the slightly higher lift/drag ratio in the DBL air vehicles. The increased drag in the train of Nimbus 3s moves the azimuth of the tether closer to $90°$ away from the heading of the ground vehicle which results in more skidding force and less propulsive force. This happens even though the improvement in lift/drag of the DBL is modest, and it emphasizes the importance of reducing drag. Since the load limit is the same, the velocity at load limit is the same, and the power is only slightly better in the DBL than in the train of Nimbus 3s. However, the attainment of maximum power occurs at a much higher velocity in the DBL (again due to high lift/drag), and consequently resulting in a much higher maximum power. The tether tension is so high, though, that even at an angular altitude below $3°$, the vertical component is sufficient to achieve lift-off of the ground vehicle before achieving maximum power. Power remains positive out to a root velocity of 239 m/s. So, if the very daunting mechanical issues of skidding, lift-off and structural failure in the air vehicles can be overcome, then a velocity well over 200 m/s, and a power well over 2 MW can be achieved by the DBL with a wind speed of only 6 m/s and with a total wing area of only 67.4 m^2.

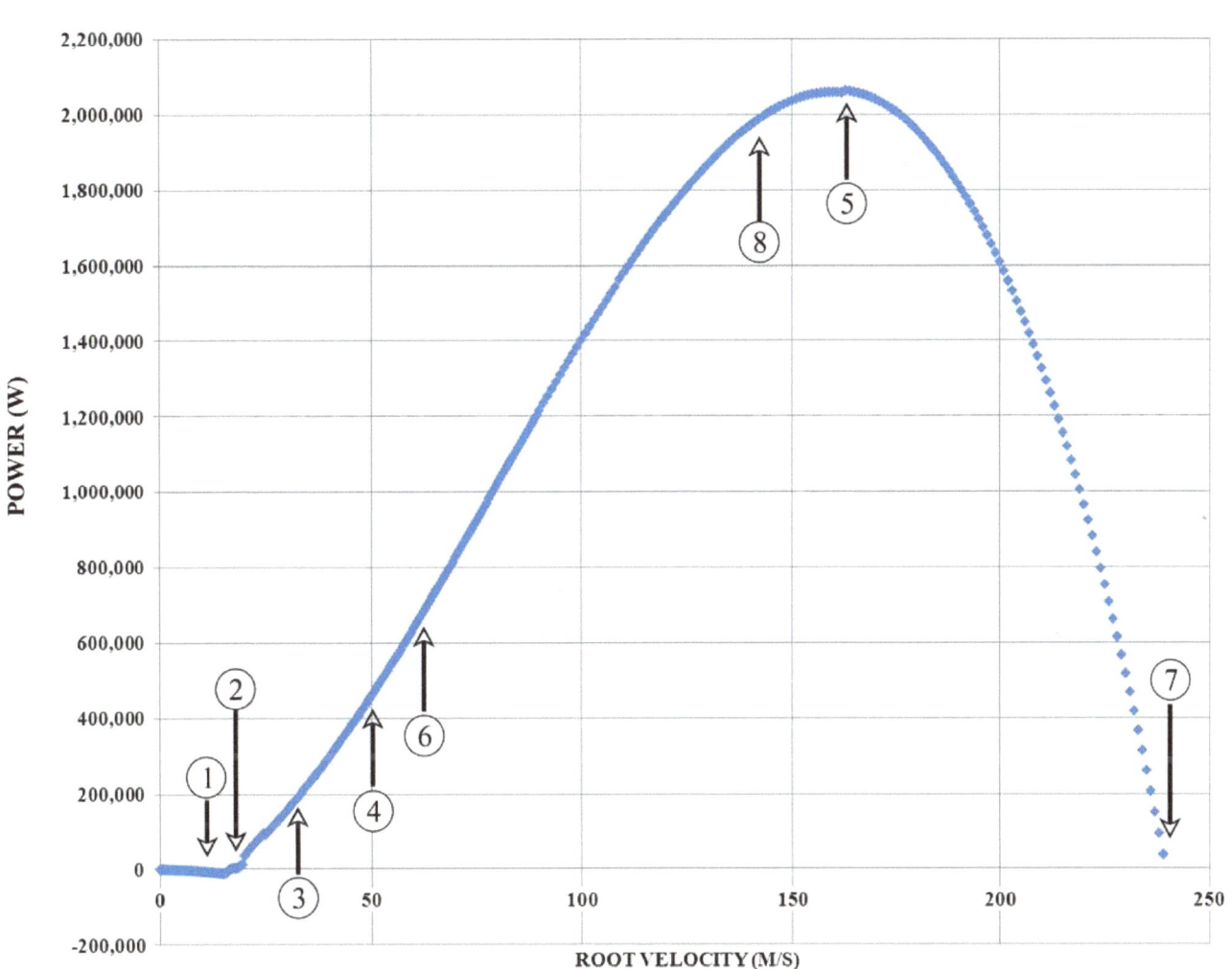

Figure 15. Power Developed by the Design Baseline (DBL) in 6 M/S Wind

Table 7. Event Markers for Figure 15

Marker	Velocity (M/S)	Heading (Deg)	Roll (Deg)	Angular Altitude (Deg)	Power (W)	Comment
1	14.5	180	0	38.3	-11,330	Air vehicle lift off. Heading is into the wind.
2	17	160	32	4.9	2,343	Power becomes positive. Propulsive traction exceeds friction and drag. Ground-vehicle engine is disengaged.
3	32	90	63	4.0	183,470	Lateral skidding force exceeds weight of ground vehicle.
4	50	90	77	3.6	458,676	About to exceed load limit of air-vehicle spar.
5	163	90	87	2.2	2,063,402	Maximum power.
6	61.5	90	81	2.8	665,094	Lateral skidding force exceeds 4 times the weight of ground vehicle
7	239	90	87	2.6	37,491	Power almost falls to zero near maximum velocity.
8	139	90	86	2.8	1,962,705	Vertical component of tether tension exceeds ground-vehicle weight.

Well, that result was for a wind of 6 m/s. What happens if the wind falls off? The EVA spreadsheet was run for the DBL and for wind speeds at 2 m/s and below. The results are shown in Figure 16, where maximum power is plotted versus true wind speed. It is seen that the DBL sunset wind is slightly below 1.2 m/s, and at a wind speed of 1.5 m/s, the DBL produces positive power beyond the velocity of the 1g skid, and slightly beyond the velocity at the load limit (50 m/s). So, with minor reductions in the angle of attack, the DBL should achieve a velocity close to that resulting in loss of power (57 m/s).

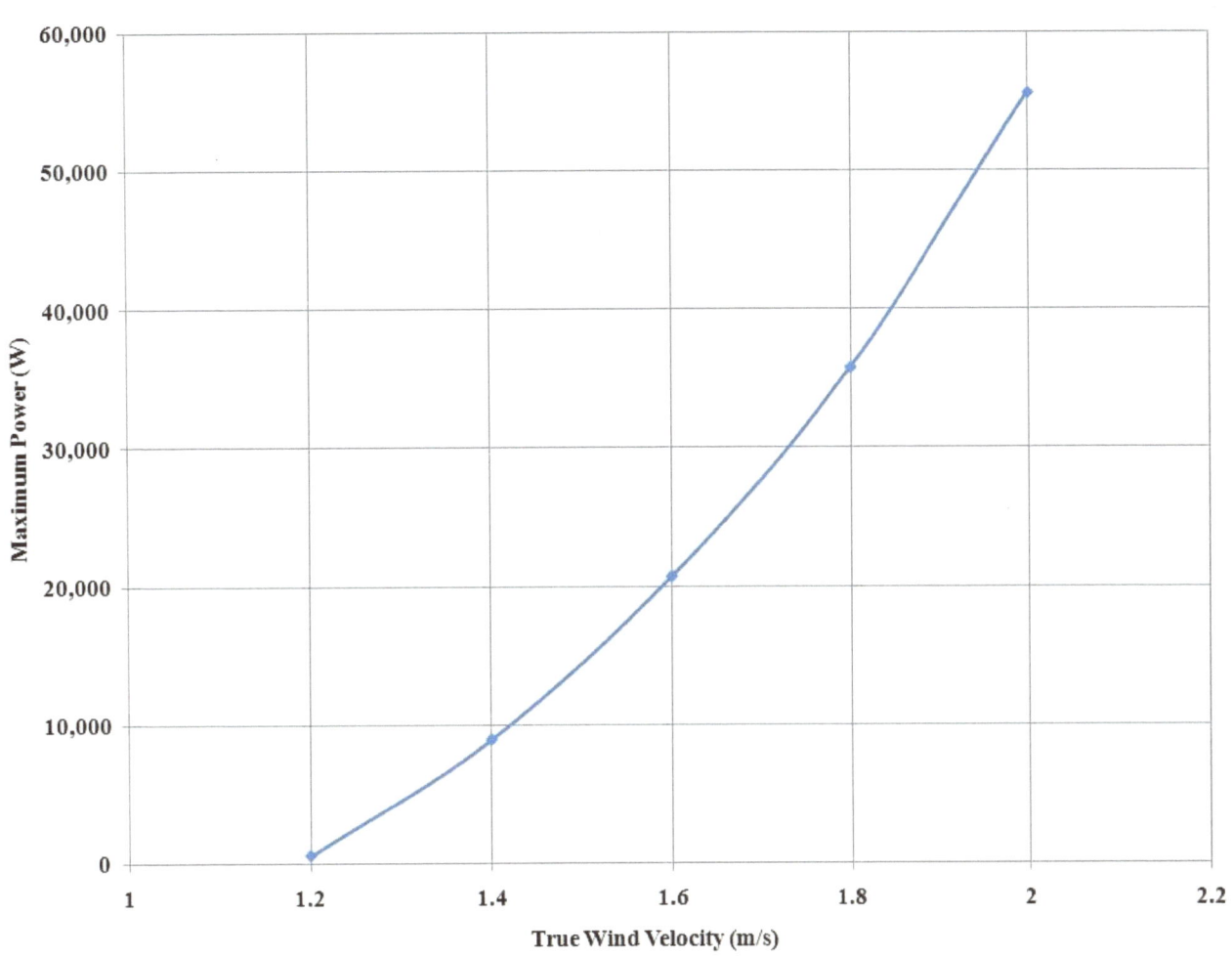

Figure 16. DBL Sunset Wind

At the other extreme, we need to know how fast the DBL will go in a 6-m/s wind. The EVA spreadsheet was run using a scheme to delay the load limit by reducing the angle of attack. The results are shown in Figure 17. Eventually, reducing the angle of attack reduces propulsive traction until it is exceeded by retardation forces, and power is lost. This occurs at a root velocity of 90 m/s (201 MPH). A similar scheme was used with the Nimbus 3 towing a Geo Metro (see Figure 4.)

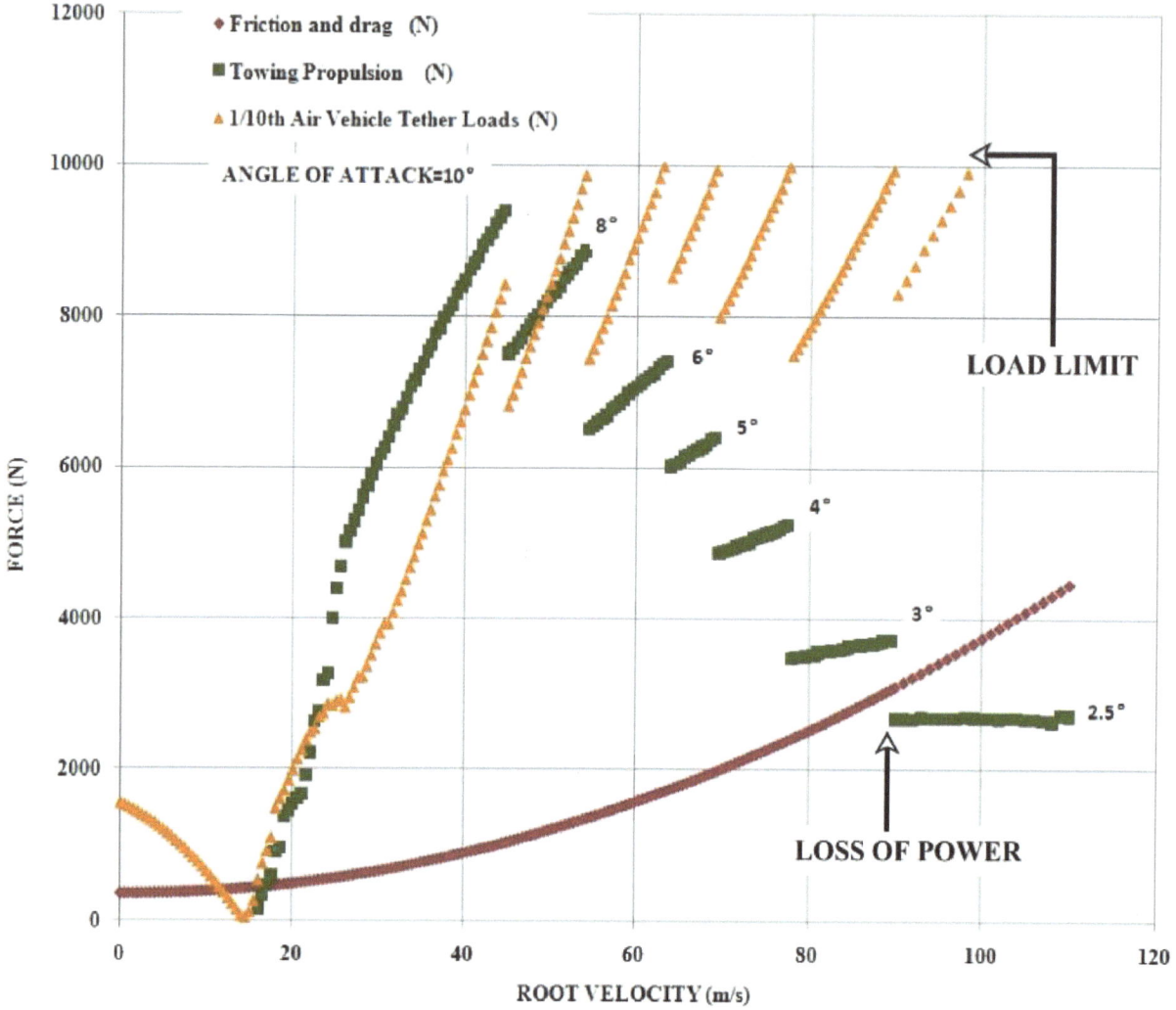

Figure 17. DBL Forces Using a Reduced Angle of Attack to Achieve Maximum Velocity

While we might expect periods of lulls in the wind, there will also be extended periods with winds higher than 6 m/s. The EVA spreadsheet was run in order to evaluate the DBL at various wind speeds ranging up to 9 m/s. The A straight line fit to the calculations shows that the power at the load limit varies with the first power of wind speed as seen in Figure 18. Similarly, the calculations show that maximum power varies with the cube of wind speed (see Figure 19.) Once again, if the daunting mechanical issues were overcome, the DBL could deliver 7 MW of power in a 9 m/s wind. The analysis of the Horizontal Axis Wind Turbine (HAWT) in Appendix G shows that the DBL, with 67.4 m^2 of wing area, generates as much power in a 6-m/s wind as up to 6½ acres of area swept out by an HAWT.

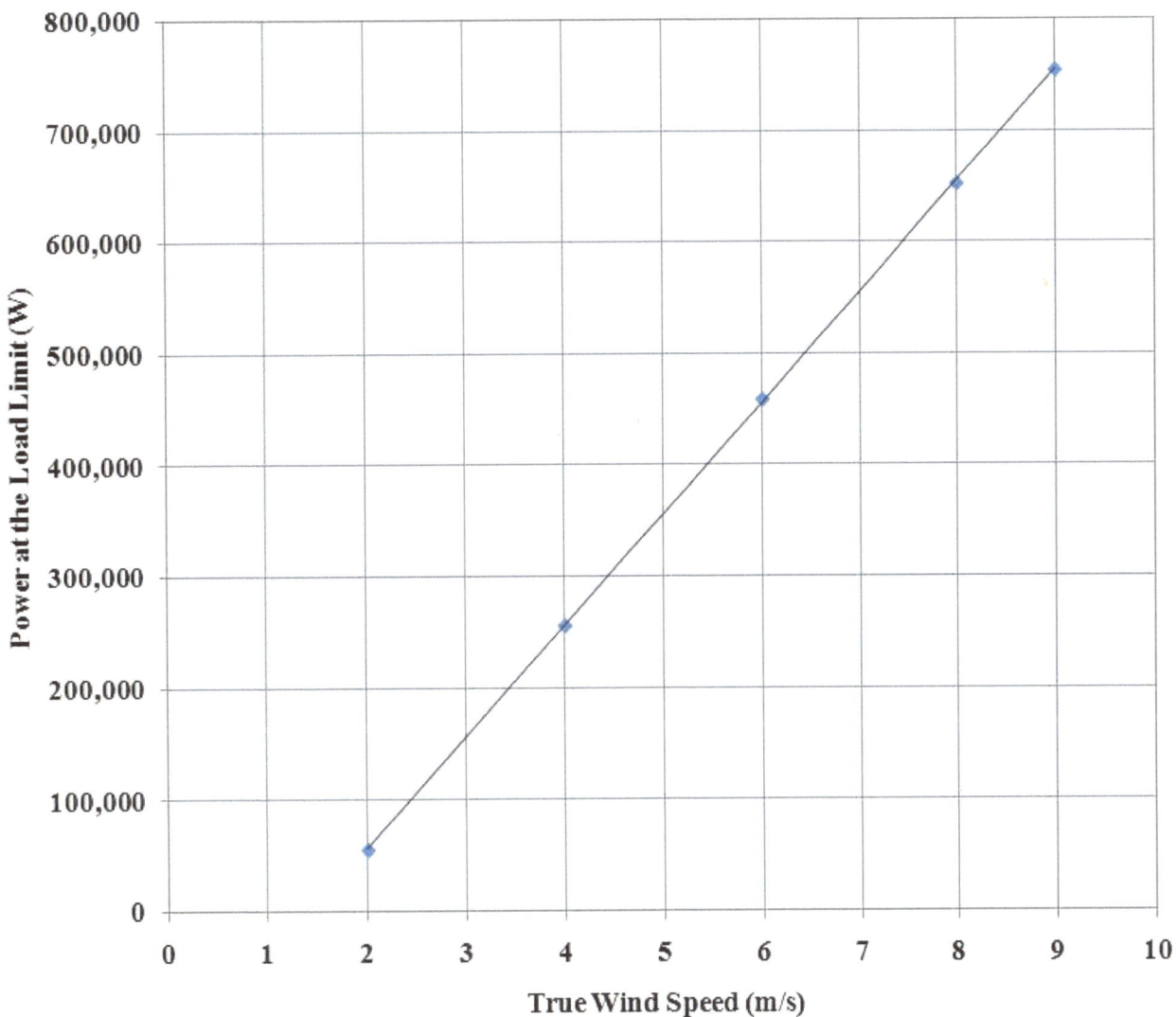

Figure 18. DBL Power at the Load Limit Versus True Wind Speed

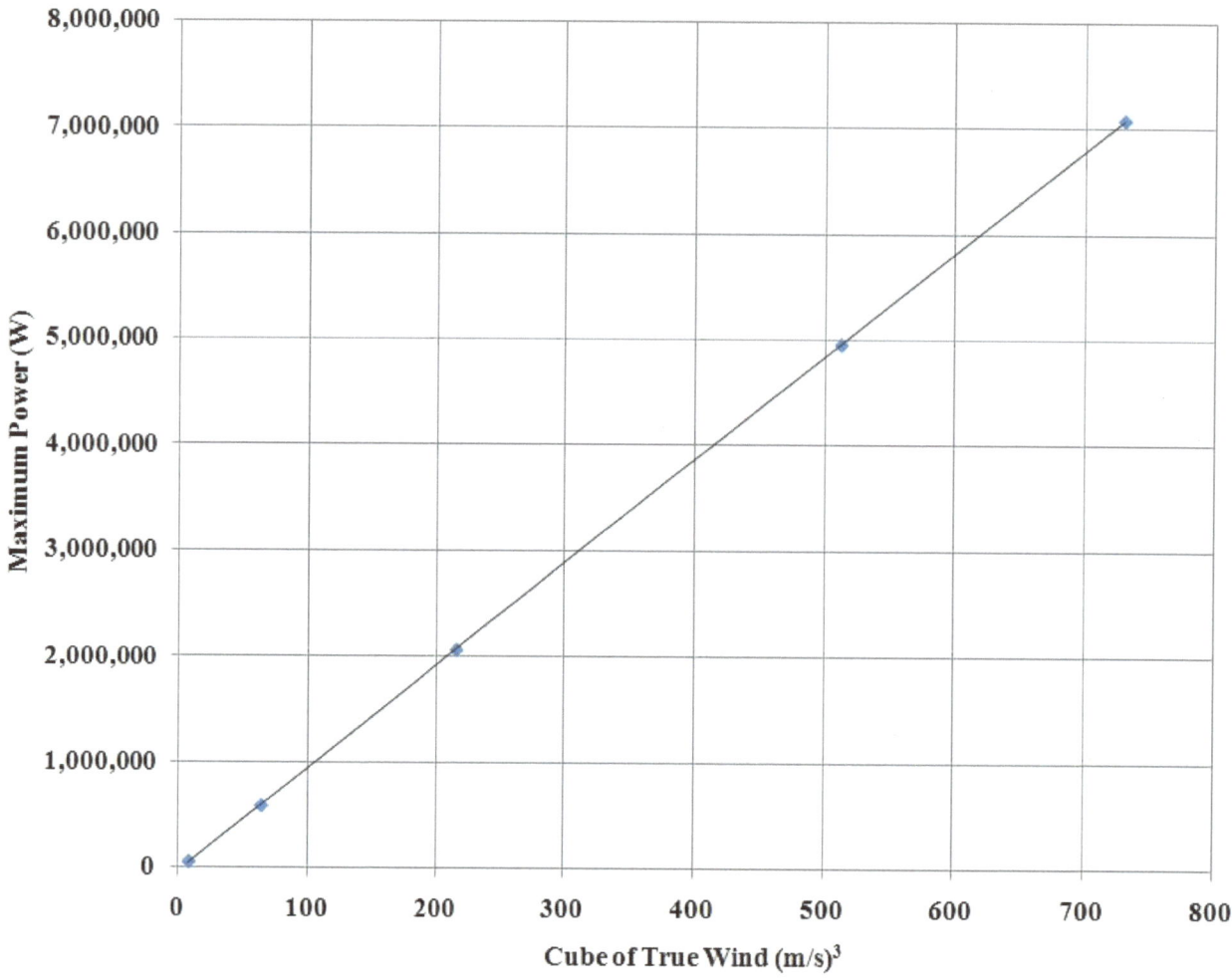

Figure 19. DBL Maximum Power Versus the Cube of the True Wind Speed

All of the calculations of power and velocity so far have been for a reach at a 90° heading. Well, what happens if we need to head the DBL in another direction? The EVA spreadsheet was run for various headings, and the results for load-limited power and velocity are shown in Figures 20 and 21. The load-limited velocity used a scheme of reducing the angle of attack, delaying the onset of load limitation as in Figures 4 and 17. Consequently, the shape of the polar plot of velocity is broader than that for power. The figures show that power is available at 20° from the upwind direction and from the downwind direction. Calculations at 15° from each direction also show usable power. But, even at 20°, it is seen that the DBL can beat into a 6-m/s wind at 55 m/s. At 20° from the downwind direction, the DBL can tack at more than ten times the speed of the wind.

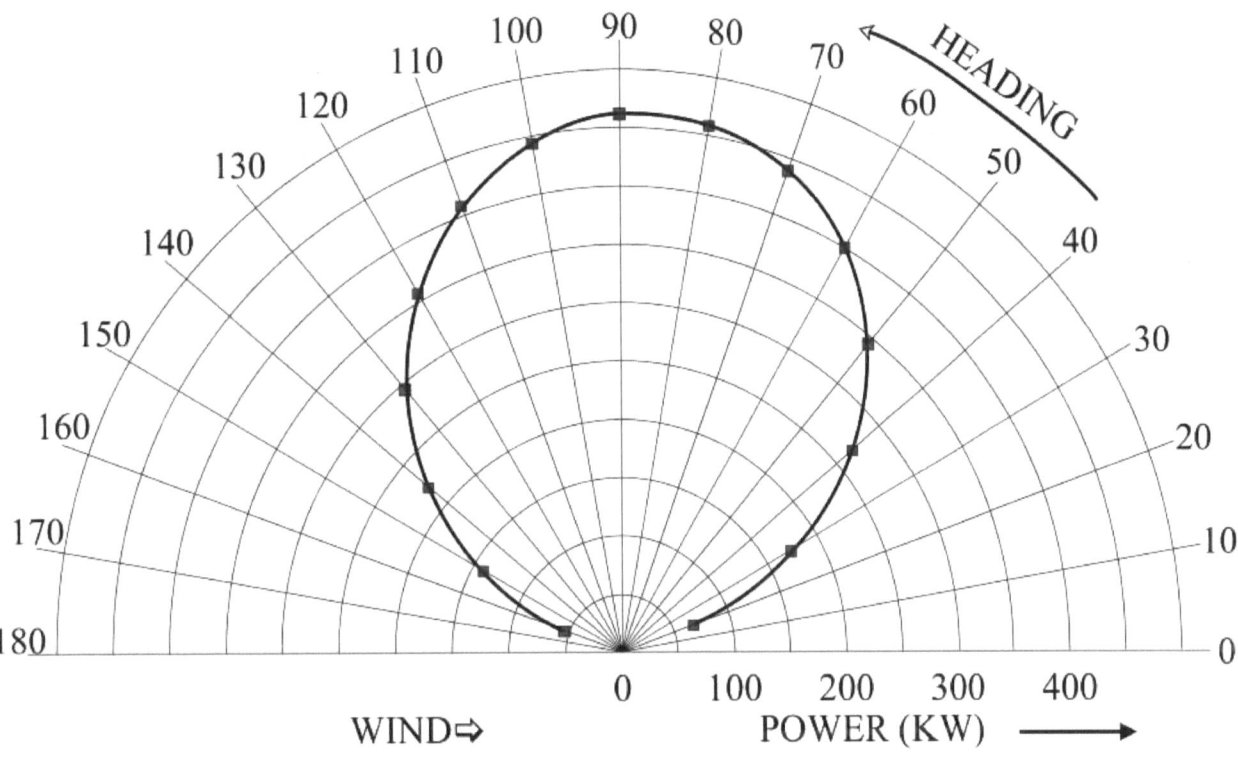

Figure 20. Polar Plot of Load-Limited Power for the DBL in a 6-m/s True Wind

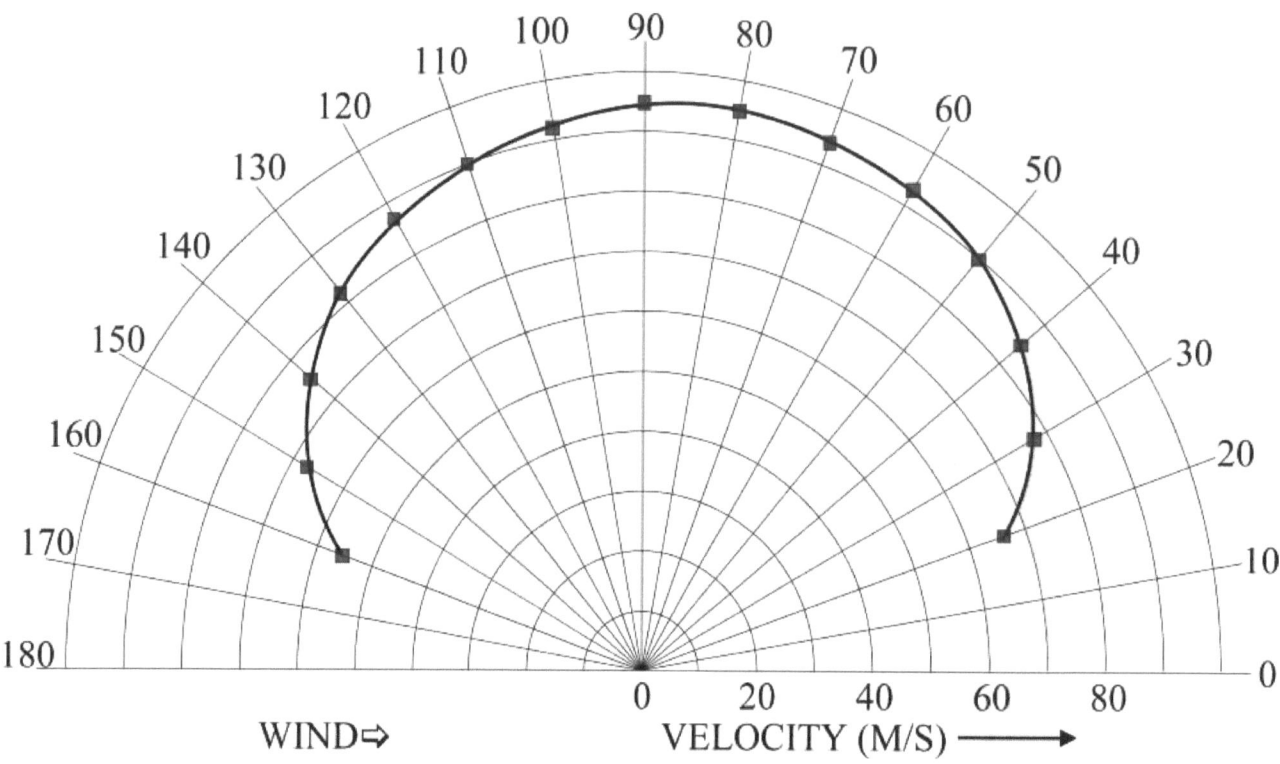

Figure 21. Polar Plot of Load-Limited Velocity for the DBL in a 6-m/s True Wind

To complete performance characterization of the design baseline, we now turn to DBL02, which consists of a train of four DBL02 air vehicles towing a DBL ground vehicle. The EVA spreadsheet was run for the DBL02 configuration, and the calculations of power are shown in Figure 22. The markers for events in this figure are the same as were used previously, and they are defined for this configuration in Table 8. Even though the DBL02 has nearly twice the wing area of the DBL, the maximum power is only moderately higher. However, the advantage in the DBL02 is in the load limit which is increased by nearly an order of magnitude over that of the DBL. Consequently, DBL02 power at the load limit (2.6 MW) is well over five times that of the DBL.

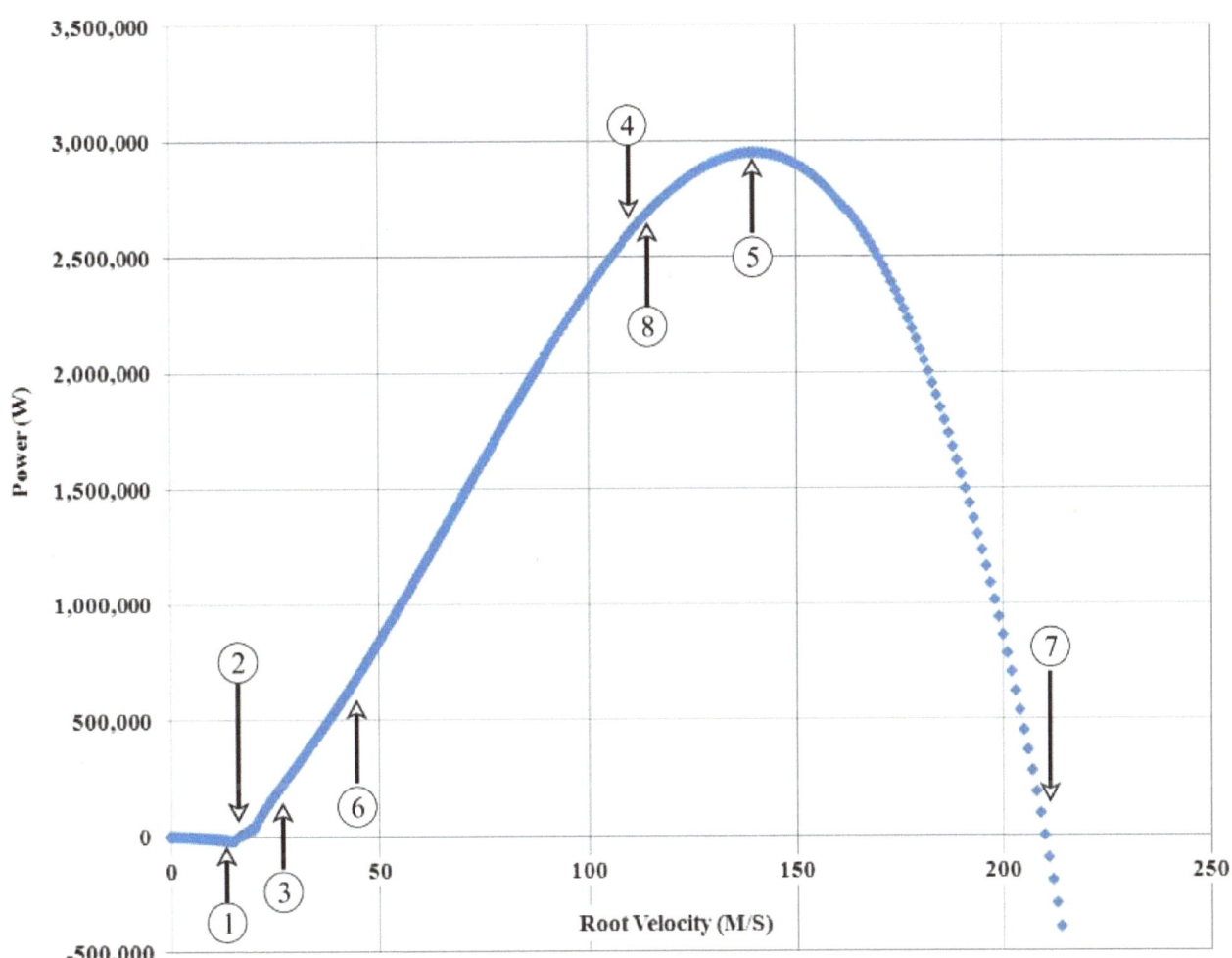

Figure 22. DBL02 Power (Markers Defined in Table 8.)

Table 8. DBL02 Event Markers

Marker	Velocity (M/S)	Heading (Deg)	Roll (Deg)	Angular Altitude (Deg)	Power (W)	Comment
1	14.5	180	0	41.69	-18,312	Air vehicle lift off. Heading is into the wind.
2	16.5	160	27	7.4	2,638	Power becomes positive. Propulsive traction exceeds friction and drag. Ground-vehicle engine is disengaged.
3	32	90	48	3.6	174,744	Lateral skidding force exceeds weight of ground vehicle.
4	111	90	86	2.1	2,622,973	About to exceed load limit of air-vehicle spar.
5	140	90	86	2.8	2,950,266	Maximum power.
6	45.5	90	76	2.8	704,731	Lateral skidding force exceeds 4 times the weight of ground vehicle
7	210	90	87	2.6	93	Power almost falls to zero near maximum velocity.
8	114	90	86	2.2	2,682,940	Vertical component of tether tension exceeds ground-vehicle weight.

The above DBL02 calculations were for a wind speed of 6 m/s. Maximum power was calculated for other speeds, and the results are shown in Figure 23. Once again, maximum power is linearly related to the cube of the wind speed. At a wind speed of 9 m/s, the maximum power is 10 MW. However, to realize that power, as before, mechanical issues would have to be solved.

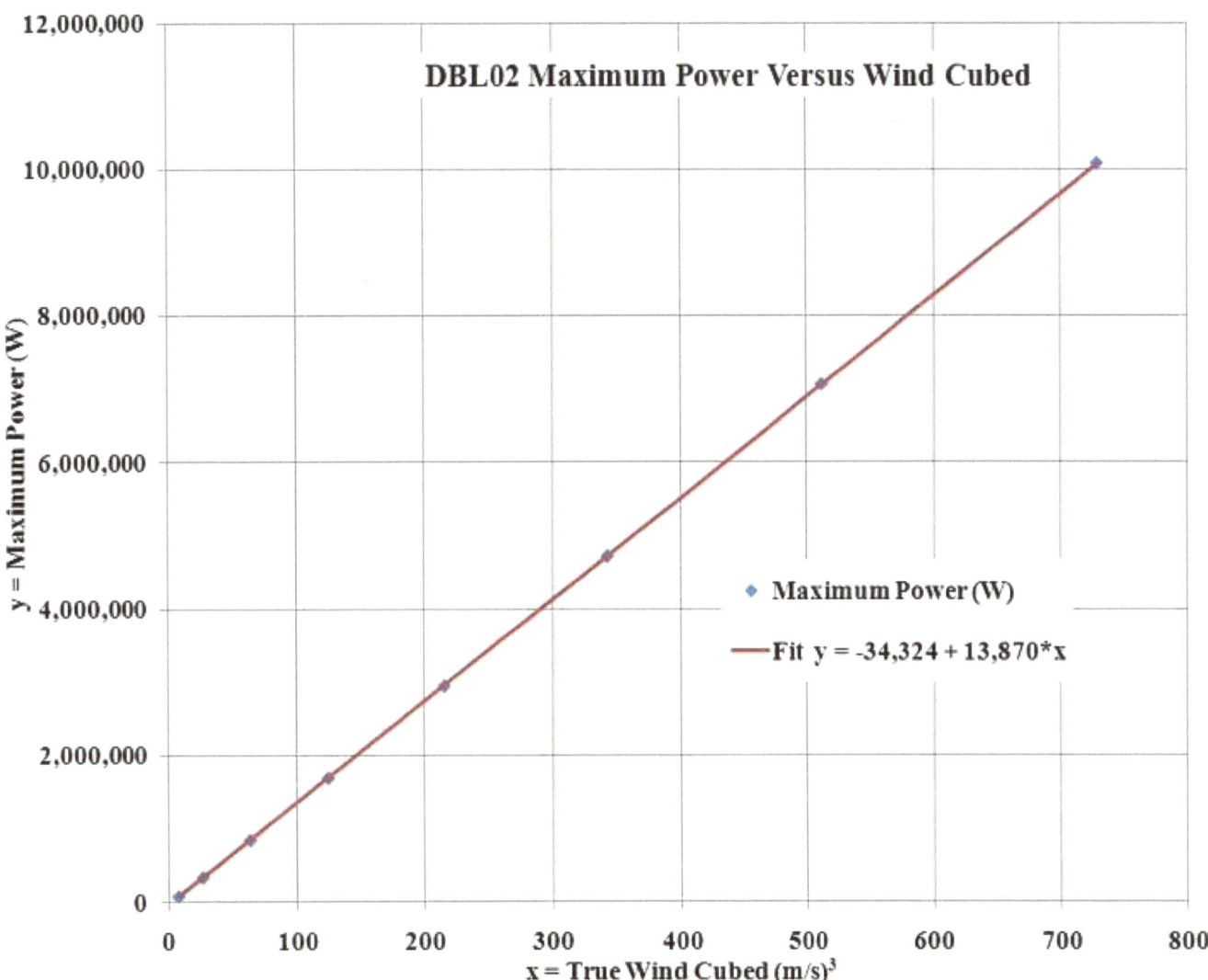

Figure 23. Linear Relationship Between DBL02 Maximum Power and Root Velocity Cubed

10. The MRK At Sea

As promised, the MRK at sea is considered. If hull drag can be accounted for in the EVA, analysis of performance should be similar to that for the land version. So, design issues and problems will be considered for the sea version, without the consideration of performance. The first issue involves the heeling of sail craft arising from aerodynamic forces on the sail centered above the center of pressure of the keel. The heeling reduces the vertical component of sail area and keel area, and directs sail lift somewhat downward. As mentioned above, the effect is reduced by the MRK, because the tether can be mounted very low, and close to the keel. But, it is not easily eliminated without mounting the tether below the waterline. Therefore, a concept referred to as the Air Keel is considered.

10.1 The Air Keel. The air keel is intended to provide an up-righting moment to counteract the heeling moment of a sail, or in the case of the MRK, a tethered sailplane. It consists of an airfoil or sail mounted vertically on a mast, set to provide lift windward. The arrangement is shown in Figures 24 and 25. At first glance, this seems to be just counter to the propelling sail or tethered sailplane, because it provides lift (horizontal and perpendicular to the apparent wind) and drag (in the same direction as the apparent wind) with components that retard the forward motion of the craft. That retardation is a consequence of the air keel, however, it is minimized by making the wing area small, and the up-righting moment arm large. Remember that the heeling of the craft is created by a moment, not a force. So, the counter to that is a moment created by a small force on a large moment arm. This will require a strong, light-weight mast that can be rotated to adjust the angle of attack of the air keel. At low boat speeds in a beam reach, the azimuth of the apparent wind will be well off of the bow, so the air keel will not be as effective or not be even needed. So, it is intended for the high boat speeds envisioned for the MRK.

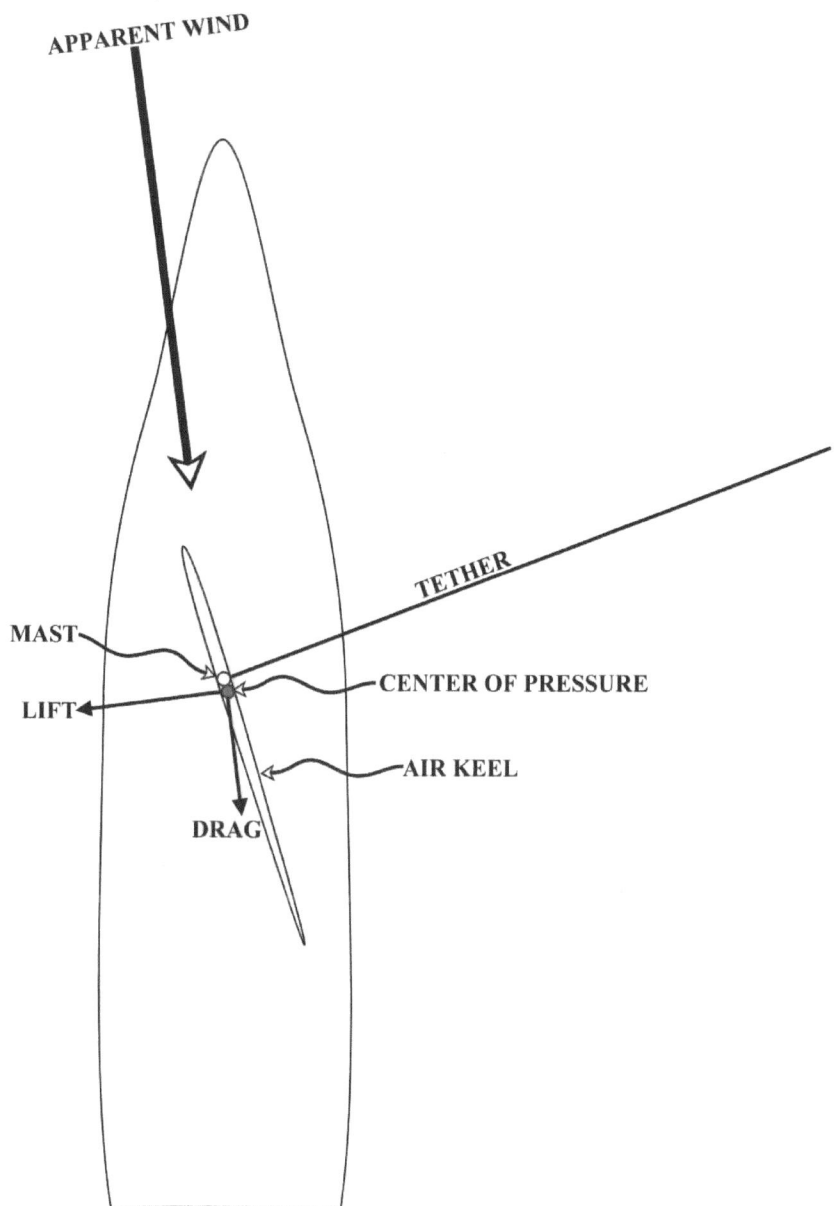

Figure 24. Top View of the Air Keel

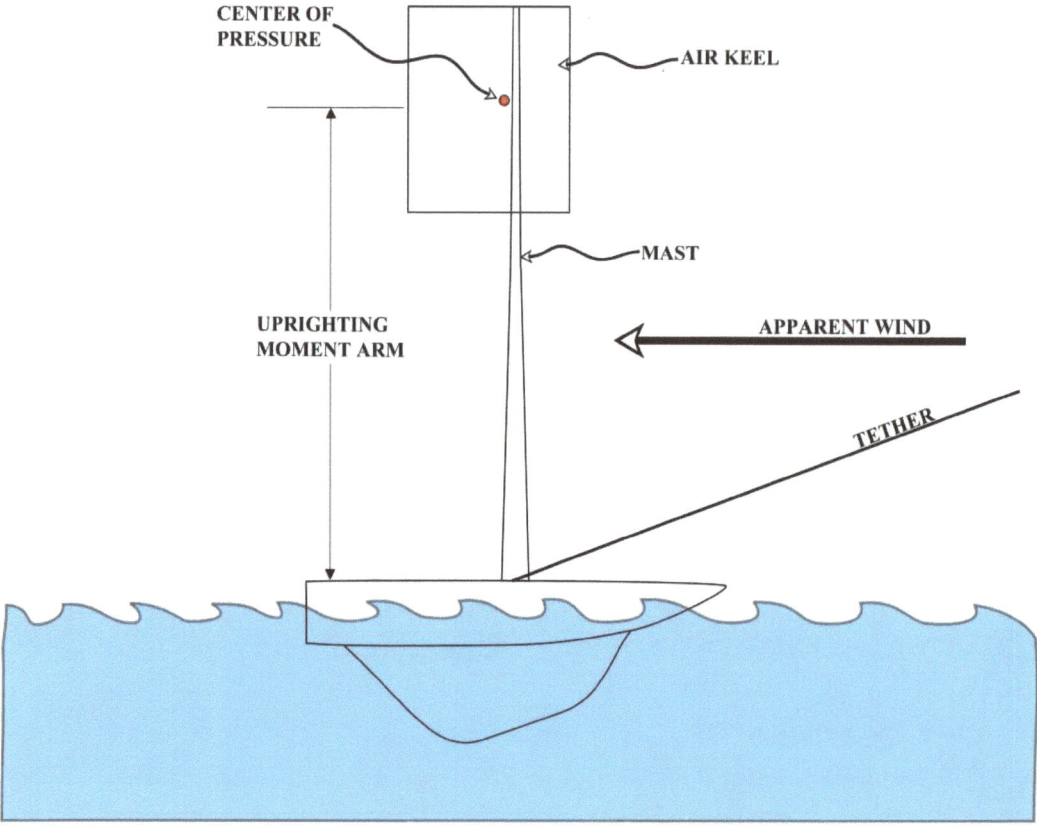

Figure 25. Side View of the Air Keel

The air keel may not be necessary if the tether can be attached to a horizontal, outboard beam, as shown in Figure 26. The figure shows the tether tension vector aligned with the center of pressure of the keel when the angular altitude of the tether is the angle ϕ. If the altitude is less than ϕ then tether tension will cause heeling alee, and if the altitude is greater than ϕ then heeling will actually be windward.

MRK HEEL AVERTMENT

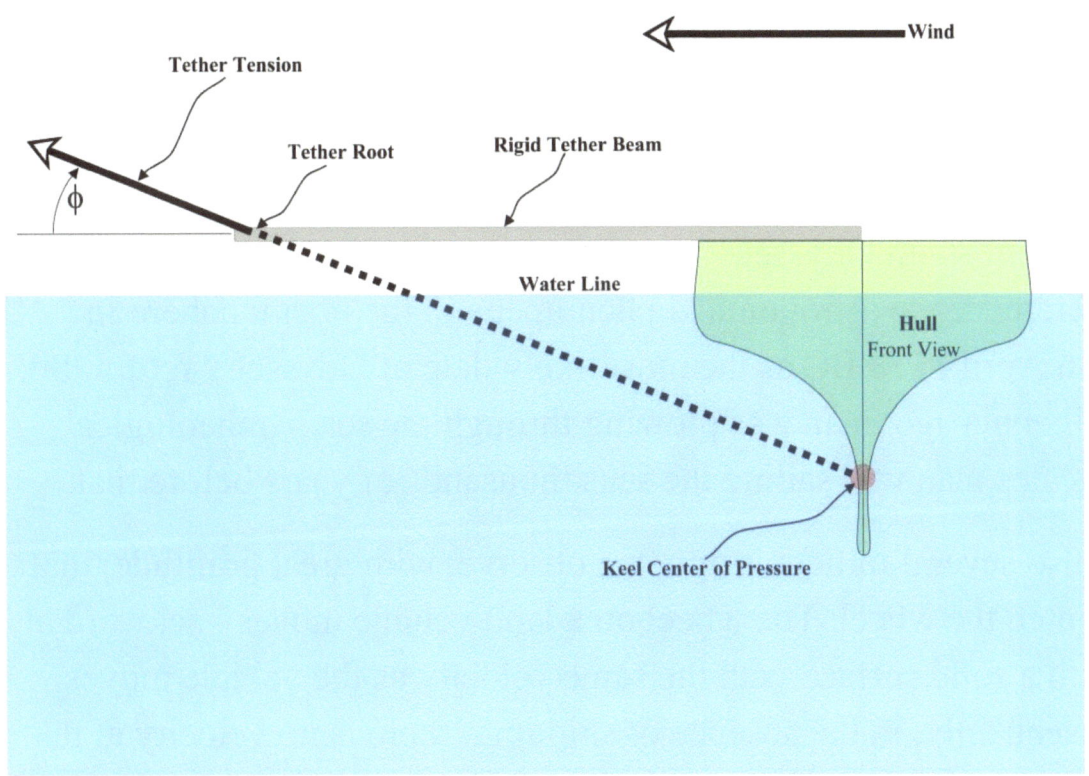

Figure 26. MRK Heel Averter

10.2 The Wheel Keel. Most sea vessels are allowed to sink down to their waterline, and then, under forward propulsion, they plow through the sea. There are some exceptions to this, such as powered hydroplaning vessels and vessels riding on an air cushion, but those vessels do not require a keel as do sailing vessels. Hydrofoil boats can raise the hull over the surface of the water, but the sea still flows over their underwater elements at the same speed as that of the boat. Furthermore, hydrofoil sail boats have not succeeded in exceeding about three times the speed of the wind. With those exceptions, transport at sea is similar to plowing a furrow into ground, which creates large retardation forces resisting forward motion. Imagine, if you will, that modern man was required to pull a plow behind him as he drove to and from his place of business. His half-hour commute would probably become two or three hours. How much time at home would he have? What sort of gasoline mileage would he get? Sounds absurd? Well, no more so than the fact that man has spent, <u>thousands</u> of years plowing furrows while moving from here to there in the sea (and still does). Homer wrote this verse in Book 12 of the Odyssey:

Next, where the Sirens dwells, you plough the seas;
Their song is death, and makes destruction please.

He wrote his poem about Odysseus' return from his journey after the Trojan War. Three references to astronomical phenomena in the poem, cohere in pinpointing 16 April 1178 BC as the most likely date of Odysseus' return. So, even three millennia ago, man was plowing through the sea. Archeological evidence indicates man was sailing the seas thousands of years before that.

Meanwhile, also several millennia ago, we observe, with great gratitude, that man had invented the wheel. The wheel on a land vehicle moves backward at its contact with a solid surface with the same velocity as the vehicle moves forward. Consequently, in the absence of skidding, there is no velocity at that point, relative to the surface. Retardation forces are greatly reduced compared to the skids and sleds and skis used in antiquity. A similar arrangement is badly needed in sea vessels. The wheel keel is intended to do that. It is especially directed toward sea vessels requiring a keel, such as sailing vessels powered by the wind.

The wheel keel consists of a bow wheel and a stern wheel with a flexible belt between them in a continuous loop. Keel elements attached to the belt extend outward (see Figures 27, 28, 29 and 30). The keel elements protrude upward above deck and downward below the water. A transverse drive fin is attached to each keel element. As the boat is propelled forward (right to left in the figures) the drive fins interact with the water which drives the belt fore to aft below water and aft to fore above deck. If the bow and stern wheels are mounted on bearings with low friction, then the effect is to move the belt and the underwater keel elements aft at nearly the same speed as the forward motion of the boat. Consequently, as with a wheel on land, there is very little motion of the keel elements relative to the surrounding water...certainly far less motion than that of the boat. As the keel elements pass around the wheels, their inner edges protrude into a groove in the wheels. This is a consequence of those inner edges assuming a chordal position within the wheels. The corresponding rotation of the elements is accommodated by pivots at their points of attachment to the belt.

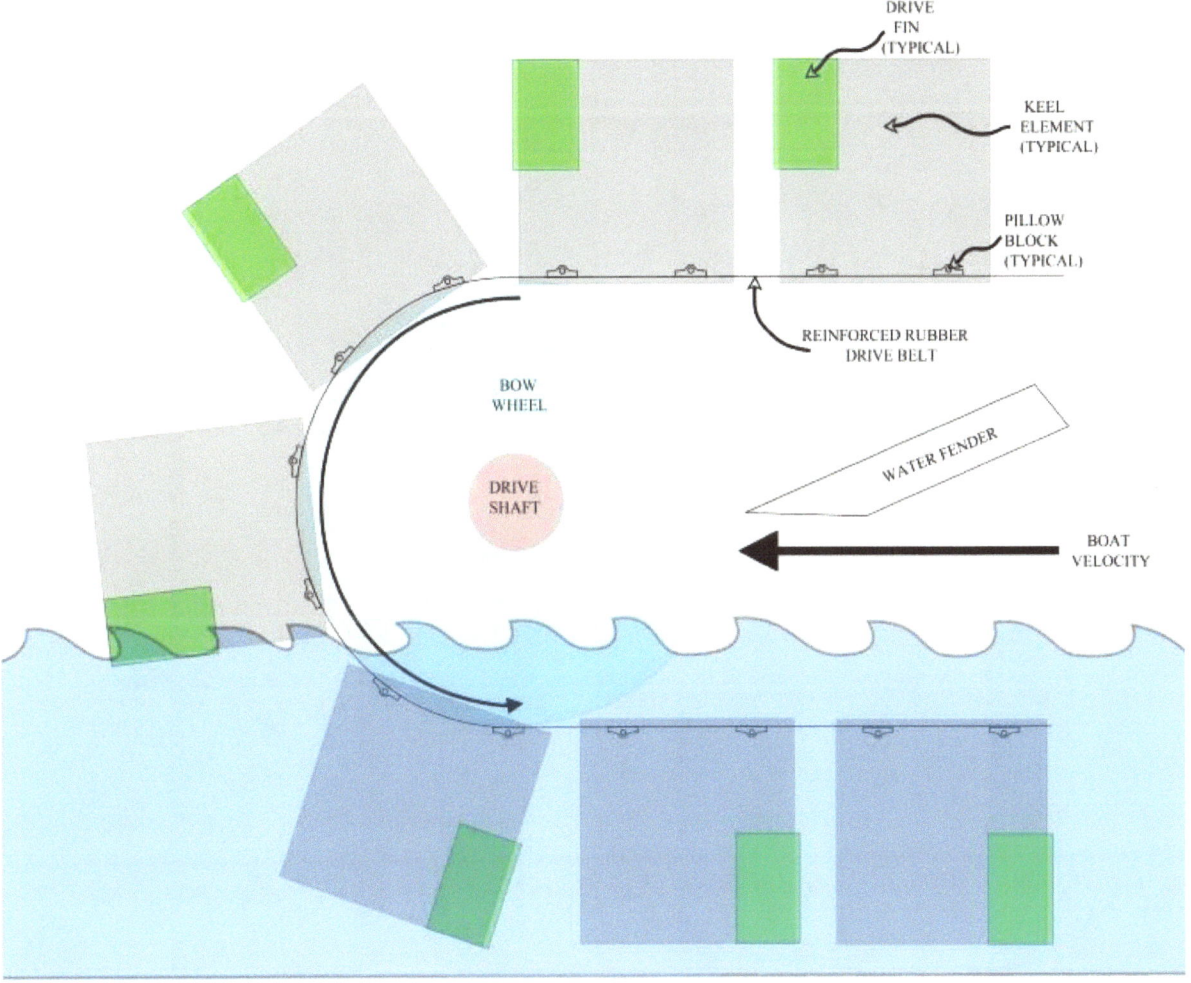

Figure 27. Bow Section of the Wheel Keel (Side View)

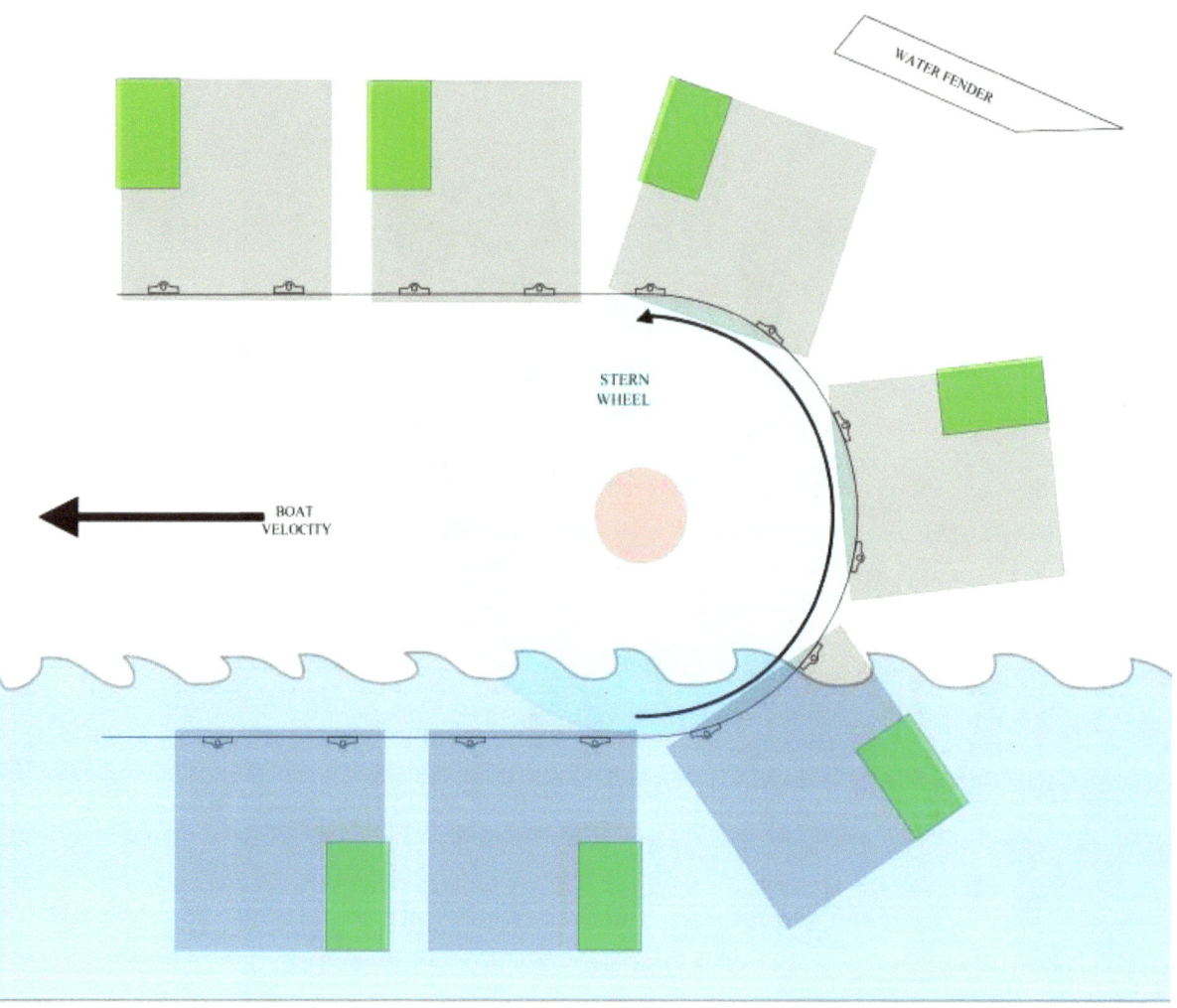

Figure 28. Stern Section of the Wheel Keel (Side View)

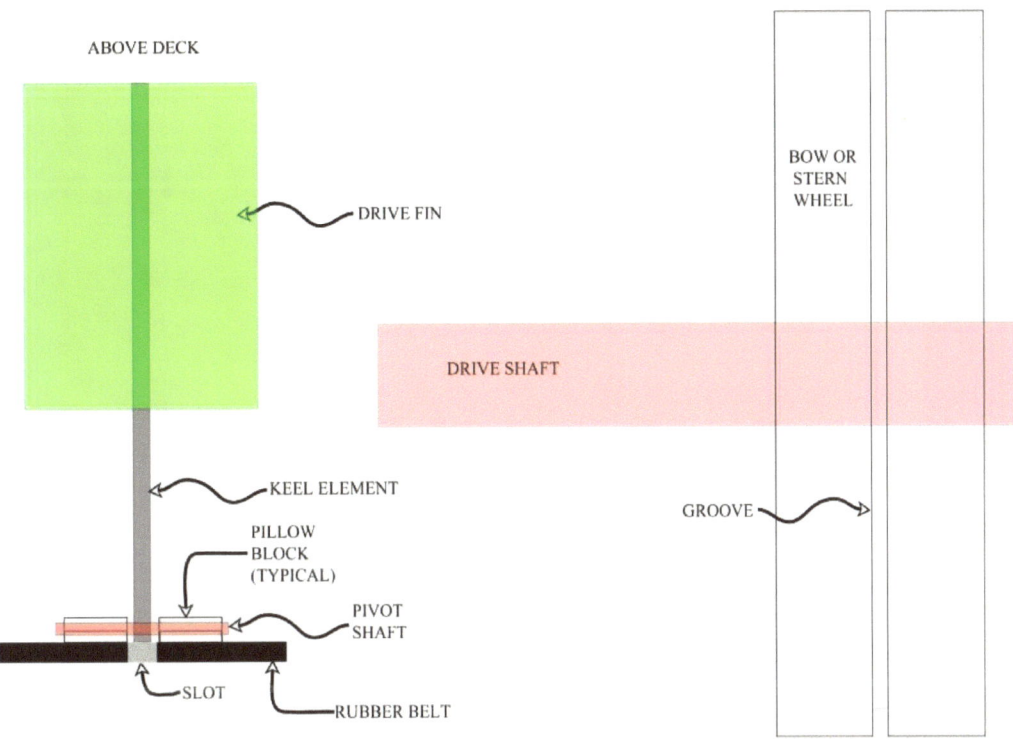

Figure 29. Keel Element and Wheel Details of the Wheel Keel (Front View)

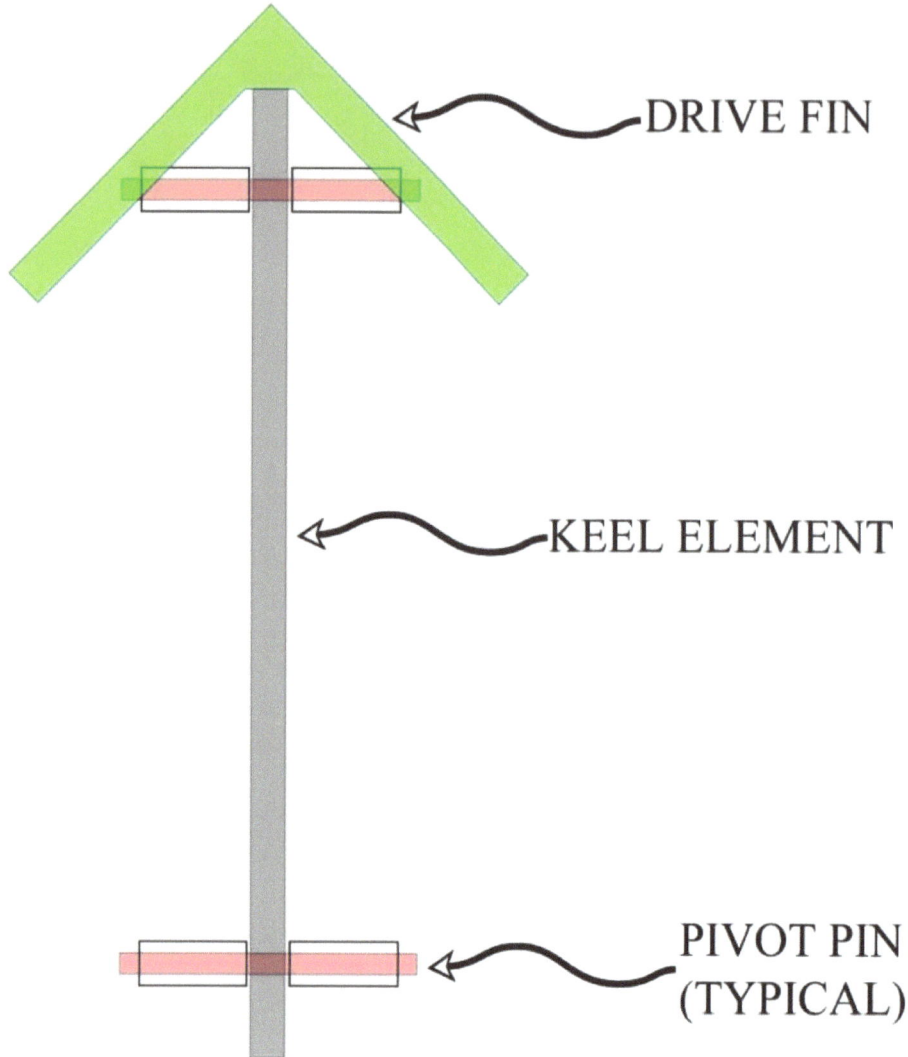

DRIVE FIN

KEEL ELEMENT

PIVOT PIN
(TYPICAL)

Figure 30. Keel Element Top View (Above Deck)

The top view of the keel element in Figure 30, shows that the drive fin has a vee shape. That is intended to reduce the impact energy as the element submerges at the bow wheel and reduces the amount of water rising with the element as it emerges from the water at the stern wheel. (One might still expect a rooster tail at the stern.) The keel element has a significant velocity relative to the water as it submerges, and again as it emerges. At these two positions, the energy exchange between the keel element and the water increases with the depth of immersion. Therefore, considering a required total keel area for the operation of the wheel-keel boat, minimizing the height of the keel element, and maximizing the distance between the bow wheel and the stern wheel would result in minimum hydrodynamic drag. At intermediate points, between the two wheels, as mentioned, there is very little velocity of the keel element relative to the water. The bow and stern wheels are shown with very little immersion in the water. It would be ideal to raise the boat until the wheels are just above the water, but with the keel elements protruding below water. This is expected to be readily done with the MRK by using the vertical component of tether tension to raise the boat. Conceivably, the mechanical power generated by the MRK system could be converted to electrical power by using one or both drive shafts to turn a generator.

10.3 Chemistry of Hydrocarbon Production Using the MRK

This primarily relates to wind energy at sea used to produce hydrogen, methane or methanol for delivery at port. The simplest concept is to generate electrical power and to use that to produce hydrogen by electrolysis of seawater. Difficulty in storage of hydrogen creates problems during production at sea and also in the logistics of consumption on land. Heavy, thick-walled storage vessels and/or extremely low-temperature cryogenics would be required. This approach should be avoided. An attractive candidate approach is to synthesize methanol as the fuel product, since it is much more easily stored and transported. This can be done as follows:

$$2H_2 + CO \rightarrow CH_3OH \text{ or}$$

$$3H_2 + CO_2 \rightarrow CH_3OH + H_2O \text{ (requires an extra mole of hydrogen per mole of methanol.)}$$

These reactions are exothermic and require a catalyst (**Cu** in **ZnO** or alumina), elevated pressure (maybe 50 atm) and moderate temperatures (250 to 500 $^\circ$C) to achieve reasonable rates. The hydrogen can be obtained as indicated above, by electrolysis:

$$2\ H_2O \rightarrow 2H^+ + 2OH^-$$

Cathode: $$2H^+ + 2c^- \rightarrow H_2$$

Anode: $2OH^- - 2e^- \rightarrow H_2O + \frac{1}{2}O_2$ The rate being improved at elevated temperatures.

The source and destination of the carbon is an important consideration. One possible approach is to pyrolyze (or reform) methanol at port for delivery of H_2 then recycling the resulting carbon by oxidation to CO or CO_2. Alternatively, the carbon can be reacted with water to form water gas:

$C + H_2O + heat \rightarrow CO + H_2$

This source of hydrogen would have to be supplemented with hydrogen from electrolysis.

Methanol appears to be a more useful and valuable product than hydrogen, because of the storage, infrastructure and transportation problems with hydrogen. But, to deliver methanol, we must have a feedstock of hydrogen, water and carbon or one of its oxides. Water is no problem at sea, and hydrogen can be supplied by electrolysis of water. But, where shall we obtain the carbon feedstock? Can we take on coal or some other source of carbon at port? Such an approach may lead to an addition of greenhouse gas into the atmosphere upon consumption of the methanol product. An attractive approach is to extract CO_2 from the atmosphere even though air is a poor feedstock for CO_2 . Using this approach, the methanol product can be consumed by combustion in air with a net neutral effect on atmospheric greenhouse gases. One might use a cryogenic process to condense dry ice from the atmosphere, which is an energy intensive process. The lime-water cycle seems to be superior:

Dissolution of limestone: $CaO + H_2O \rightarrow Ca(OH)_2$

Aeration of lime-water: $Ca(OH)_2 + CO_2 \rightarrow CaCO_3$ (precipitate) $+ H_2O$.

Calcine chalk: $CaCO_3 + heat \rightarrow CaO + CO_2$.

The best process for the production of methanol appears, at this time, to be the use of gaseous electrolysis of steam and CO_2:

$2H_2O + heat \rightarrow 2O^{--} + 4H^{+}$

$CO_2 + heat \rightarrow CO^{++} + O^{--}$

Anode: $3O^{--} -6e^{-} \rightarrow 1\frac{1}{2}O_2$ effluent

Cathode: $4H^{+} + CO^{++} + 6e^{-} \rightarrow 2H_2 + CO$

Product: $2H_2 + CO$ + pressure + moderate heat + catalyst (Cu + ZnO in $Al_2O_3) \rightarrow CH_3OH$

It is expected that the CO_2 feedstock can be obtained from industries generating CO_2 as a waste effluent. This would be a less expensive alternative to sequestration. Alternatively, if the methanol product is consumed in fuel cells or by combustion, the CO_2 effluent could be captured and returned into the above process.

Methanol is somewhat corrosive to many storage materials, and it has less energy content than gasoline. Therefore, it may be desirable to process it further. The Methanol to Gasoline process was developed by Mobil in the early 1970's. The methanol vapor is sent to a dimethyl ether (DME) reactor containing a dehydration catalyst (alumina) where approximately 75% of the methanol is dehydrated to an equilibrium mixture of DME, water and methanol.

$2CH_3OH \longleftrightarrow CH_3OCH_3 + H_2O$

The reaction is rapid, reversible and exothermic. The DME can then be further dehydrated to higher molecular weight alkanes, even gasoline.

The technology of chemically storing the energy generated by the MRK system is wide open. The variety of approaches can be summarized by paraphrasing Ms. Browning:

How do I synthesize thee? Let me count the ways.

11. Future Work

11.1 Acceleration Vector Analysis. As mentioned above, the technique of sheeting involves a systematic change in the velocity of the air vehicles relative to the surface vehicle, which has the effect of increasing the average air speed of the air vehicles. This produces more lift, and more power delivered by the air vehicles without increasing the speed or the corresponding drag of the surface vehicle. But the EVA did not consider changes in velocity and the associated accelerations. The evaluation of the benefits of sheeting to the MRK is one reason to conduct an acceleration vector analysis.

The other benefit involves accelerations of the surface and air vehicles. Sailboats typically achieve their highest water speed traveling about perpendicular to the wind, which is referred to as a beam reach. The apparent wind is off the windward bow, and the keel resists transverse aerodynamic forces which would otherwise cause a leeward drift. The transverse force is usually higher than the component of force propelling the vessel forward. If the heading is adjusted somewhat leeward, the boat would initially accelerate until the leeward drift reduces the apparent wind. It is suggested that the initial acceleration be used to increase average speed, and then, when the apparent wind falls off, the heading be returned to the reach. This interval is akin to a leap followed by a reach.

One of the things to be done in the future is to outline and analyze a leaping reach, used to increase the average speed of an MRK. The leaping reach has particular application when a sailboat is propelled by a kite, because the kite can provide a vertical component of tether tension which tends to lift the boat out of the water. In this case, the leap can be air born, reducing drag, and the term *leaping* reach attains added significance. Then the reach can be resumed after the vessel alights. This is commonly done by kite surfers, or kite boarders. With some experience, they can manage to alight very smoothly and gently. Such a vessel would skip across the surface, much like a skipping stone.

11.2 Air Vehicle Improvements. MRK power and velocity was shown to be limited by the load limit of the spar and/or transverse forces leading to skidding or lift off. Improvements in those parameters should be attainable in future designs. A load factor of greater than 40 was achieved by DS aircraft with a span around 3 meters and an aspect ratio below 14. The present analysis assumed a linear relationship between the load factor and the aspect ratio in an attempt to find an optimum balance between those parameters. There are two uncertainties here. One is that assumed relationship, and the other is that the load factor of 40 is a minimum. What load factor can be attained for a larger scale, unmanned vehicle? That needs to be addressed.

This study shows the very significant effect of air-vehicle drag on performance. The design of unmanned tethered aircraft needs to be optimized to reduce drag and weight. These parameters might be improved if the slave vehicles in a train do not require horizontal and vertical stabilizers and control surfaces. Their attitude may be controlled by multiple tethers to the master air vehicle under radio control.

11.3 Ground Vehicle Resistance to Transverse Forces. Presuming that reasonable care is taken in the design of the ground vehicle for the friction-like rolling resistance and the coefficient of aerodynamic drag (the latter being more important), the significant issues center around resistance to transverse forces causing skidding and lift off. That resistance can be increased by increasing the weight of the ground vehicle. That will also increase the friction-like rolling resistance. An alternative is to adapt an air foil to the ground vehicle to provide an aerodynamic down force, increasing resistance to both skidding and lift off. But, that will increase both rolling resistance and aerodynamic drag. So, increasing weight seems to be superior.

The best solution from a technical standpoint appears to be mounting the surface vehicle on a rail or a fixed cable. If the rail structure were mechanically capable of resisting tether forces, the MRK systems would be capable of reaching theoretical maxima in power and velocity. Those maxima are extremely large. But, the rail approach would require a large capital investment.

The polar for DBL maximum power is not shown here. However, it shows that an array of three rails forming an equilateral triangle would always provide above 60% of maximum power at any wind direction.

The Israeli firm, Planum Vision LTD, has been promoting a concept they call the Train Cable Unmanned Air Vehicle (TCUVA), intended for surveillance. The TCUAV consists of a UAV tethered to a train car. The train car rides on a pair of over-head cables which provides electrical power to the car and, via a conducting tether, to the UAV. (See http://www.planumvision.com/ and http://thefutureofthings.com/news/1055/tcuav-an-unmanned-aerial-surveillance-system.html) This system is exactly what is needed for the MRK, EXCEPT that the TCUAV rail car provides electrical power to the UAV and the MRK air vehicle would provide mechanical power to the rail car. The wheels of the rail car could turn a generator for the generation of electrical power. Alternatively, the rail-car wheels could be the magnetic rotor of a generator. As the rim of the wheel passes by the cable, it would generate electrical power in the cable. Perhaps Planum Vision LTD could be convinced to use the TCUAV for the generation of power. The concept could work at sea as well. Then, lateral traction would be provided by a cable supported by anchored towers which do not move through the sea. That would eliminate hull drag in a surface vessel moving through the sea. Once again, it would be very capital intensive.

12. Conclusions

The land MRK produces very large amounts of power compared to other methods of harvesting power from the wind. The DBL version produces almost ½ MW at the load limit, with a maximum velocity of 90 m/s. This with only 67.4 m^2 of wing area in a wind of only 6 m/s. The DBL02 version produces more than 2½ MW at the load limit (much higher velocity) with a wing area of 125 m^2. If the limitations of spar load limit, skidding and lift off are removed (as with a rail), then the DBL would deliver more than 2 MW, which is the equivalent of the power delivered by an HAWT array sweeping out a total area of 6½ acres. The maximum power for a DBL02 would be almost 3 MW. In a not-unusual 9-m/s wind, the DBL would deliver 7 MW and the DBL02 would deliver 10 MW.

A sea version of the MRK would take advantage of better wind conditions. The available area at sea is very large. The oceans have a total area of 335 million square kilometers (129 million square miles), and subtracting polar ice caps and territorial waters, about half of that would be unobstructed, readily available and appropriate for the MRK. In comparison, the "wind corridor" in the United States running from the Canadian border to Texas, and between the Rocky Mountains and the Mississippi River adds up to about 2 million square kilometers (0.77 million square miles). This corridor is highly obstructed with limited access for an MRK. The wind energy resources in the ocean are vastly superior to those on land, and no one will complain about visual pollution when the MRK is hundreds or even thousands of kilometers offshore. But, the MRK at sea would have to deal with the conversion of all that mechanical power into electrical power, storing the electrical power in the form of a hydrocarbon product, and the transport of the product to shore. If a wheel keel is used, the wheel drive shafts can turn a generator, so there would be no need for wind turbines. The best candidate for the chemical storage product is methanol. A tanker used to transport the methanol to shore would arrive with the required carbon dioxide feedstock.

Well...we have come upon a concept for harvesting more power from the wind than any other known method, and a concept for wind-powered travel faster than any other known method. All it took was some freshman physics, a little calculus and matrix algebra, throw in a pinch of elementary chemistry and a layman's view of aerodynamics. Put all of these ingredients into a caldron, stir well, and out of the witch's brew emerges Project Sea Tree. Each of the steps along the way are very easily within the capabilities of many, many people. So, considering that the motive is substantial, why has it not been done yet? That is a mystery. There is much more work needed to reduce this concept to practice. All of those BYLBs out there are needed. This writing has made liberal use of the plural form of the first and second person. It is time to wander into the first person singular. I am a fairly old man, and I need more than four hours of sleep every night. But, the BYLBs don't need more sleep than that. So, all of you BYLBs: uproot your tethers, weigh your anchors, and let's get moving. The rewards for your labors will be astounding. PERIOD.

FINI

Dennis Wakefield Stevens
May, 2011

<u>Appendix A. Equilibrium Vector Analysis Of The Moving-Root Kite</u>

The extension of the kite problem to allow motion of the tether root is addressed here. The object is to demonstrate the amount of power that the kite or tethered air vessel can deliver to the surface vessel. The analysis uses the alt-azimuth coordinate system which employs the azimuthal angle, θ, the altitude, ϕ, and the radius, r, as shown in Figure 1.

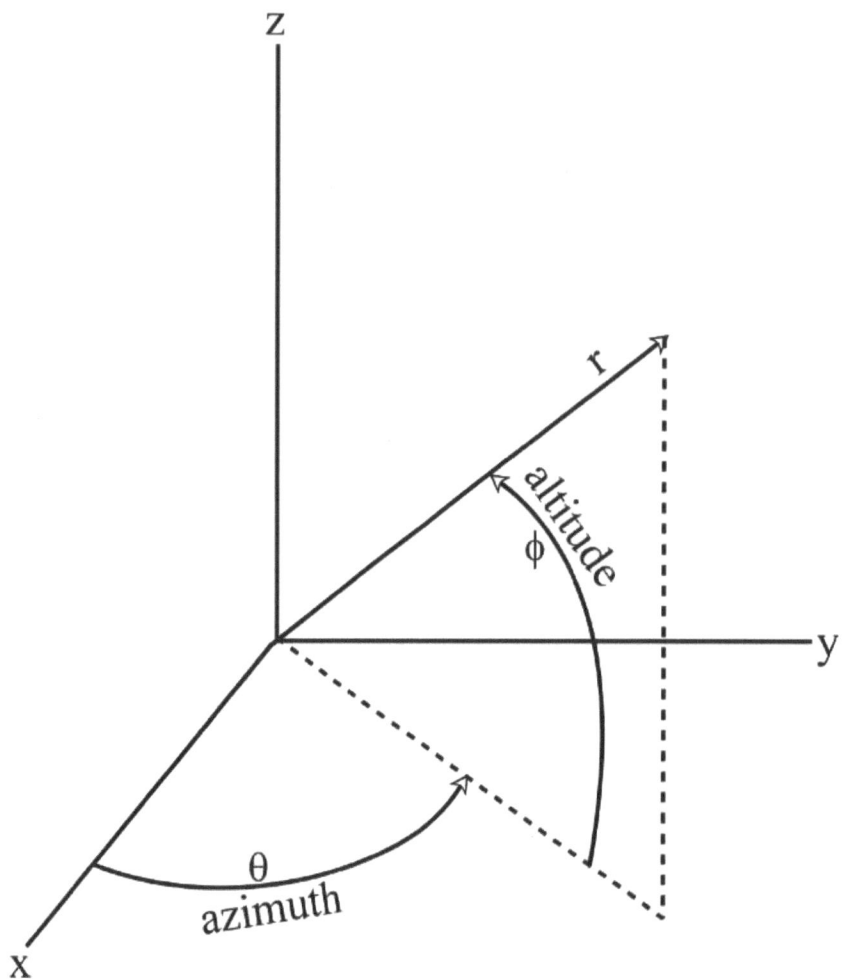

FIGURE 1. ALT-AZIMUTH COORDINATE SYSTEM

Vector components in the three Cartesian coordinates will be expressed in alt-azimuth coordinates. However, the vectors themselves will be expressed in terms of the x-y-z unit vectors. No attempt is made to use the alt-azimuth unit vectors. In this analysis. the surface vessel is at the origin and the tethered sailplane is at the tip of the $\vec{r}$ vector. Since the sailplane is constrained by the tether, r is a constant equal to the length of the tether, and does not enter into the analysis.

Figure 2 shows the vectors parallel to the surface of the earth, with wind, $\vec{V_w}$, parallel to the x-axis, and surface vessel velocity, $\vec{V_s}$, parallel to the surface of the earth at an azimuth θ_S. The resultant apparent wind velocity, $\vec{V}_A$, and aerodynamic drag on the air vehicle, $\vec{D}$, at azimuth θ_A are also parallel to the surface of the earth. The projection of the tether onto the surface of the earth is at azimuth θ_{TW}, relative to the true wind. Note that the zero azimuthal angle was chosen to be down wind in contrast to conventions used in the literature. Most polar plots of sailboat velocity use the up-wind direction as the zero azimuth. That would be less desirable in this analysis, because it would result in a true-wind vector in direct opposition to the x-axis.

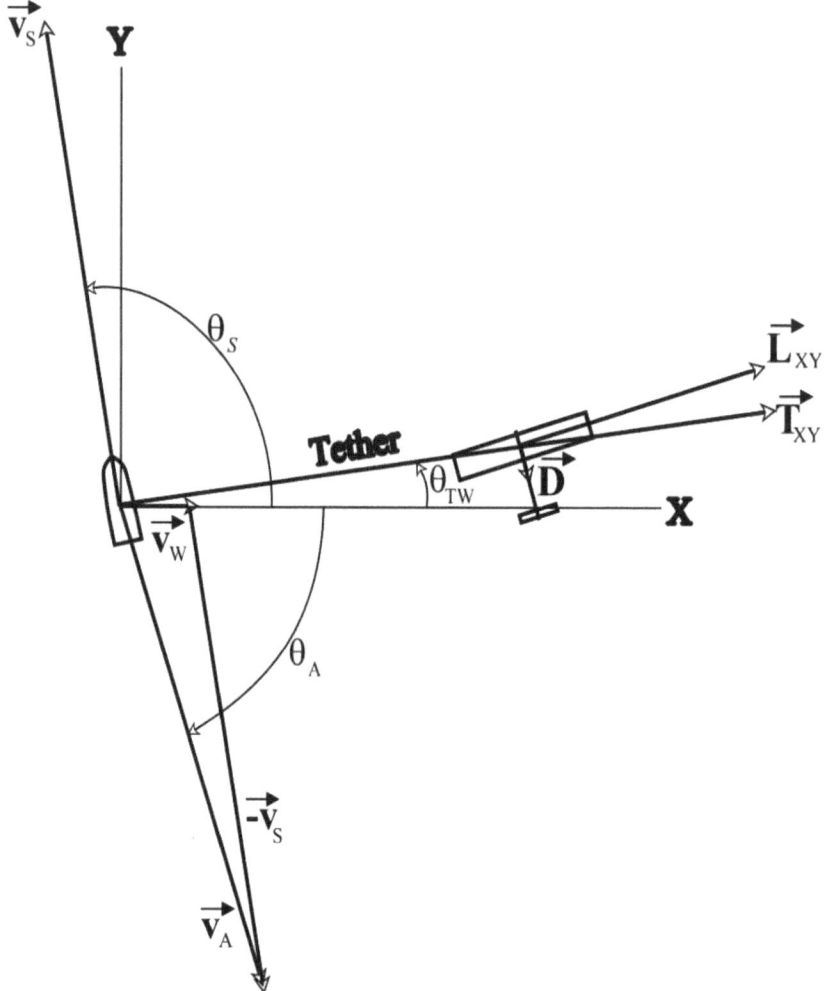

**FIGURE 2. VECTOR DIAGRAM
LOOKING DOWN ON THE X-Y PLANE**

The velocity of the surface vessel is

1.) $\vec{v_S} = v_S \begin{bmatrix} \cos(\theta_S) \\ \sin(\theta_S) \\ 0 \end{bmatrix}$

where the square brackets contain the x- y- and z-components of the vector. Its magnitude, v_S, and azimuth are set by the operator.

The wind velocity apparent at the air vehicle, and the surface vessel is shown in Figure 2 as the vector difference between the true wind velocity and the surface-vessel velocity:

$$\vec{v_A} = \vec{v_W} - \vec{v_S} = \begin{bmatrix} v_W - v_S \cos(\theta_S) \\ -v_S \sin(\theta_S) \\ 0 \end{bmatrix}.$$

The magnitude of the apparent wind is v_A,

2.), $v_A = \sqrt{v_W^2 - 2v_W v_S \cos(\theta_S) + v_S^2}$

and its azimuth, θ_A, is given by

3.), $\cos(\theta_A) = \dfrac{v_W - v_S \cos(\theta_S)}{v_A}$

and

4.) $\sin(\theta_A) = -\dfrac{v_S}{v_A} \sin(\theta_S)$

Note that the tether-root motion could be facilitated by attaching it to a moving ship at sea. It could also be attached to a wheeled vehicle on land or to a vehicle with blades on ice. The air vehicle could be a sailplane with enhanced performance characteristics resulting from its high-airspeed capabilities and its high lift-drag characteristics. The sailplane would be kept aloft by the apparent wind. The force vectors on the sailplane include weight, $\vec{W}$, and the aerodynamic forces, lift, $\vec{L}$, and drag, $\vec{D}$. These forces are reacted by tension in the tether, $\vec{T}$, to produce mechanical equilibrium. Drag and the horizontal component of, $\vec{T}$ ($\vec{T}_{XY}$), are shown in Figure 2.

This analysis assumes a sailplane with aerodynamic control surfaces set by the operator for angle of attack, α, roll, β, and perhaps yaw, although a vertical stabilizer should hold the heading to the same azimuth as the apparent wind. Figure 3 shows the vertical plane containing the weight vector, the wing span, the lift vector and roll, and Figure 4 shows the angular altitude of the tether and the angle of attack of the air vessel.

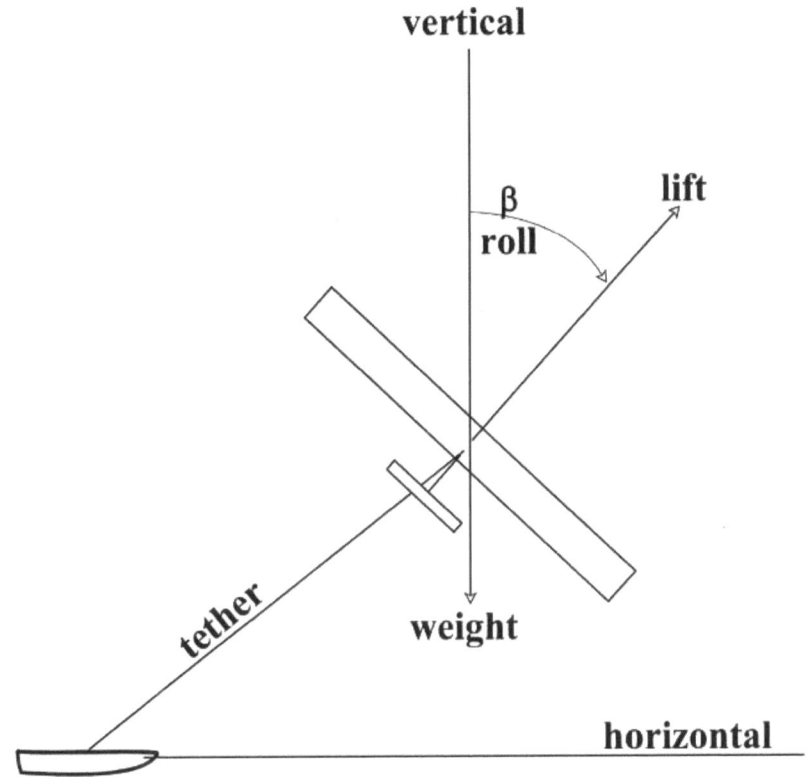

**FIGURE 3. VERICAL PLANE CONTAINING LIFT,
ROLL WEIGHT AND WING SPAN**

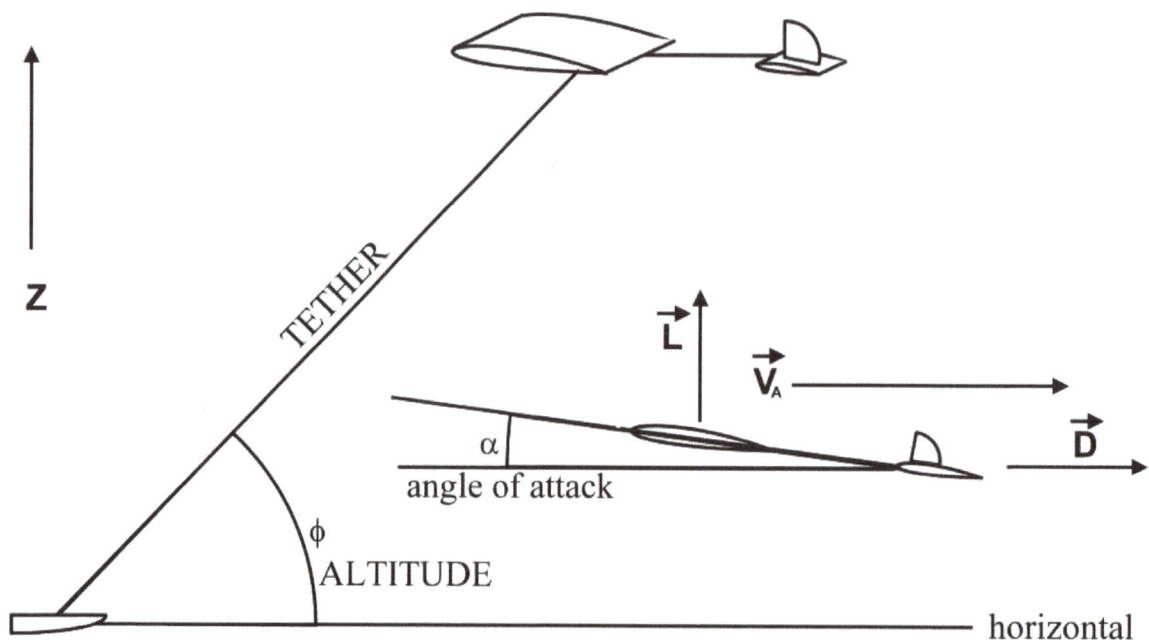

FIGURE 4. VERTICAL PLANE CONTAINING THE TETHER AND TILTED PLANE CONTAINING LIFT, DRAG AND APPARENT WIND VELOCITY

The force vectors on the sailplane are:

$$5.) \quad \vec{W} = W \begin{bmatrix} 0 \\ 0 \\ -1 \end{bmatrix}$$

$$6.) \quad \vec{L} = L \begin{bmatrix} -\sin(\beta)\sin(\theta_A) \\ \sin(\beta)\cos(\theta_A) \\ \cos(\beta) \end{bmatrix}$$

and

7.) $\vec{D} = D \begin{bmatrix} \cos(\theta_A) \\ \sin(\theta_A) \\ 0 \end{bmatrix}$

The magnitudes of these vectors are scalar weight of the sailplane, W, (Tether weight is not addressed here.) scalar lift, L, and scalar drag, D, where:

8.) $L = 0.5\rho_a V_a^2 S C_L$

and

9.) $D = 0.5\rho_a V_a^2 S C_D$

In these equations, ρ_a, is air density, S is the wing area, C_L is the coefficient of lift and C_D is the coefficient of drag. Note that $L/D = C_L/C_D$. The coefficients of lift and drag are highly dependent upon design of the air vessel. However, data on common airfoil performance leads to the following first approximations taken to demonstrate the potential of this approach:

$C_L = \alpha/10$, where α is in degrees.

$C_D = 0.009 + C_L^2/(3A)$, where A is the aspect ratio of the wing. Most of these calculations will use $\alpha = 10°$ ($C_L = 1$) to avoid flow separation.

Tether tension, $\vec{T}$, in reaction to the vectors , $\vec{W}$. $\vec{L}$ and $\vec{D}$ is given by:

$\vec{T} = \vec{W} + \vec{L} + \vec{D}$, and from 5.), 6.) and 7.):

10.) $\vec{T} = \begin{bmatrix} D\cos(\theta_A) - L\sin(\beta)\sin(\theta_A) \\ D\sin(\theta_A) + L\sin(\beta)\cos(\theta_A) \\ L\cos(\beta) - W \end{bmatrix}$

The azimuth of the apparent wind, θ_A, is eliminated by substituting 3.) and 4.):

$$\vec{T} = \begin{bmatrix} \dfrac{D}{V_A}(v_W - v_S\cos(\theta_S)) + L\dfrac{V_S}{V_A}\sin(\beta)\sin(\theta_S) \\[3mm] -D\dfrac{V_S}{V_A}\sin(\theta_S) + \dfrac{L}{V_A}\sin(\beta)(v_W - v_S\cos(\theta_S)) \\[3mm] L\cos(\beta) - W \end{bmatrix} \equiv \begin{bmatrix} T_x \\ T_y \\ T_z \end{bmatrix}$$

which has magnitude

$$T = \sqrt{T_x^2 + T_y^2 + T_z^2} = \sqrt{D^2 + L^2 + W^2 - 2LW\cos(\beta)}$$

The azimuth of the tether relative to the true wind is θ_{TW}, which is given by:

$$\tan(\theta_{TW}) = T_y/T_x \,.$$

This result yields the tether tension vector in the coordinate system with the true wind vector along the x-axis. To get the result relative to the surface vessel, requires rotation of the tether tension vector about the z-axis. The rotation matrix for the generalized rotation angle, θ, is:

$$R_z(\theta) = \begin{bmatrix} \cos(\theta) & -\sin(\theta) & 0 \\ \sin(\theta) & \cos(\theta) & 0 \\ 0 & 0 & 1 \end{bmatrix}$$

The requirement is for clockwise rotation through the angle θ_S as viewed at positive altitude looking down on the horizontal x-y plane. The alt-azimuth coordinate system provides for positive azimuth when the angle is generated by counterclockwise rotation (see Figure 1.). Therefore, the desired rotation angle is $-\theta_S$. So, the rotation matrix for the desired rotation is:

$$R_z(-\theta_S) = \begin{bmatrix} \cos(\theta_S) & \sin(\theta_S) & 0 \\ -\sin(\theta_S) & \cos(\theta_S) & 0 \\ 0 & 0 & 1 \end{bmatrix}$$

With the heading of the surface vessel rotated to the x-axis, $\vec{T}$ becomes:

$$\vec{T_S} = R_z(-\theta_S)\vec{T}$$

$$= \begin{bmatrix} \cos(\theta_S) & \sin(\theta_S) & 0 \\ -\sin(\theta_S) & \cos(\theta_S) & 0 \\ 0 & 0 & 1 \end{bmatrix} \begin{bmatrix} \dfrac{D}{v_A}(v_W - v_S\cos(\theta_S)) + L\dfrac{v_S}{v_A}\sin(\beta)\sin(\theta_S) \\ -D\dfrac{v_S}{v_A}\sin(\theta_S) + \dfrac{L}{v_A}\sin(\beta)(v_W - v_S\cos(\theta_S)) \\ L\cos(\beta) - W \end{bmatrix}$$

$$= \begin{bmatrix} \dfrac{D}{v_A}\cos(\theta_S)(v_W - v_S\cos(\theta_S)) + L\dfrac{v_S}{v_A}\sin(\beta)\sin(\theta_S)\cos(\theta_S) - D\dfrac{v_S}{v_A}\sin^2(\theta_S) + \dfrac{L}{v_A}\sin(\beta)\sin(\theta_S)(v_W - v_S\cos(\theta_S)) \\ \dfrac{-D}{v_A}\sin(\theta_S)(v_W - v_S\cos(\theta_S)) - L\dfrac{v_S}{v_A}\sin(\beta)\sin^2(\theta_S) - D\dfrac{v_S}{v_A}\cos(\theta_S)\sin(\theta_S) + \dfrac{L}{v_A}\sin(\beta)\cos(\theta_S)(v_W - v_S\cos(\theta_S)) \\ L\cos(\beta) - W \end{bmatrix}$$

Collecting terms in D and in L, and simplifying yields:

$$11.)\quad \vec{T_S} = \begin{bmatrix} \dfrac{D}{v_A}(v_W\cos(\theta_S) - v_S) + L\dfrac{v_W}{v_A}\sin(\beta)\sin(\theta_S) \\ -D\dfrac{v_W}{v_A}\sin(\theta_S) + \dfrac{L}{v_A}\sin(\beta)(v_W\cos(\theta_S) - v_S) \\ L\cos(\beta) - W \end{bmatrix} \equiv \begin{bmatrix} T_p \\ T_t \\ T_z \end{bmatrix}$$

Here, tether tension has a propulsion component along the heading of the surface vessel, T_p, a horizontal component transverse to the heading of the surface vessel, T_t, and a vertical component, T_z.

Rotating $\vec{v_s}$ through the same angle changes $\vec{v_s}$ to

$$v_s \begin{bmatrix} 1 \\ 0 \\ 0 \end{bmatrix}.$$

Power, P, delivered to the surface vessel is obtained by taking the dot product of this vector with $\vec{T_s}$:

$$P = v_s \begin{bmatrix} 1 \\ 0 \\ 0 \end{bmatrix} \cdot \begin{bmatrix} \dfrac{D}{V_A}(v_W\cos(\theta_S)-v_s)+L\dfrac{V_W}{V_A}\sin(\beta)\sin(\theta_S) \\ -D\dfrac{V_W}{V_A}\sin(\theta_S)+\dfrac{L}{V_A}\sin(\beta)(v_W\cos(\theta_S)-v_s) \\ L\cos(\beta)-W \end{bmatrix}$$

This leads to:

12.) $\quad P = v_s\left(\dfrac{D}{V_A}(v_W\cos(\theta_S)-v_s)+L\dfrac{V_W}{V_A}\sin(\beta)\sin(\theta_S)\right)$

Net usable power can be obtained by subtracting off power wasted in moving the surface vessel against resistive forces such as drag and/or friction.

The vertical component of the tether tension is:

13.) $\quad T_z = L\cos(\beta)-W$

and the angular altitude, ϕ, is obtained from:

14.) $\sin(\phi) = T_z/T$

The lateral force opposed by the keel in a sea vessel and by skid resistant wheels in a ground vehicle is:

15.) $\quad T_t = -D\dfrac{V_W}{V_A}\sin(\theta_S) + \dfrac{L}{V_A}\sin(\beta)(v_W\cos(\theta_S) - v_S)$

The azimuth of the tether relative to the surface vessel heading is θ_{TS}, which is given by:

16.) $\quad \tan(\theta_{TS}) = T_t / T_p$

(See equation 11.)

These relationships are in terms of what would be known quantities or parameters set by the operator, and provide all the information required to determine power delivered from the air vehicle to the surface vessel.

Appendix B. Maximum Lift/Drag Ratio

The Moving Root vector analysis for a sailplane tethered to a surface vehicle uses a simplification of the lift and drag coefficients for a feasibility analysis. The assumed simplifications in these coefficients are:

1.) $C_L = \alpha/10$

and

2.) $C_D = C_{D,0} + \dfrac{C_L^2}{3A}$

where α is the angle of attack expressed in degrees, $C_{D,0}$ is the value of C_D at zero lift (typically 0.009), and A is the aspect ratio of the wing. The analysis shows that the lift/drag ratio (L/D) is equal to C_L/C_D. So, as α and C_L are raised above zero, L/D initially rises, being dominated by C_L . But eventually drag in the denominator takes over, and L/D falls off. This indicates that L/D passes through a critical value (maximum) as α and C_L are increased. The specifications for a sailplane usually provide the aspect ratio and the maximum L/D (or the equivalent maximum glide ratio). However, $C_{D,0}$ is not usually given. For example, the specifications for the MDM-1 Fox aerobatic sailplane (i.e. capable of high-g maneuvers) provide A = 15.93 and a maximum L/D of 28. But $C_{D,0}$ is not given. The objective of this analysis is to use 1.) and 2.) above to derive $C_{D,0}$.

The L/D ratio is the ratio of 1.) to 2,) :

$$L/D = C_L/C_D = \dfrac{C_L}{C_{D,0} + \dfrac{C_L^2}{3A}}$$

It is convenient to redefine the independent and dependent variables as:

$Y = C_L / C_D$, and

$X = C_L$, leading to:

3.) $$Y = \frac{X}{C_{D,0} + \dfrac{X^2}{3A}}$$

Bringing the denominator to the left side,

$$Y\left(C_{D,0} + \frac{X^2}{3A}\right) = X$$

Differentiating this with respect to X,

$$Y'\left(C_{D,0} + \frac{X^2}{3A}\right) + \frac{2XY}{3A} = 1$$, where

$$Y' = dY/dX .$$

The critical value of Y, (Y_{CR}) is obtained by setting Y' = 0. This occurs at the critical value of X, (X_{CR}). Then

$$\frac{2 X_{CR} Y_{CR}}{3A} = 1 .$$

Using 3.) at criticality, this becomes

$$\frac{\dfrac{2 X_{CR}^2}{3A}}{C_{D,0} + \dfrac{X_{CR}^2}{3A}} = 1 .$$

Moving the denominator on the left to the right side

$$\frac{2\,X_{CR}^2}{3\,A} = C_{D,0} + \frac{X_{CR}^2}{3\,A} \text{, and} \quad C_{D,0} = \frac{X_{CR}^2}{3\,A}.$$

Solving for X_{CR},

4.) $\quad X_{CR} = \sqrt{3\,A\,C_{D,0}}$

and inserting this into 3.)

$$Y_{CR} = \frac{\sqrt{3\,A C_{D,0}}}{C_{D,0} + \frac{3\,A C_{D,0}}{3\,A}} \text{, or}$$

$$Y_{CR} = \sqrt{\frac{3}{4}\frac{A}{C_{D,0}}}.$$

Since the objective is to get $C_{D,0}$ given A and Y_{CR}, solve this for $C_{D,0}$

5.) $\quad C_{D,0} = \frac{3}{4}\frac{A}{Y_{CR}^2}$

Inserting 5.) into 4.)

6.) $\quad X_{CR} = \frac{3}{2}\frac{A}{Y_{CR}}$

The example given at the beginning of this analysis was for the MDM-1 Fox sailplane. From its specified aspect ratio, 15.93, and maximum lift-drag ratio, 28, equations 5.) and 6.) yield $C_{D,0} = 0.01524$ and $X_{CR} = 0.8537$. Using these values in 3.) , lift/drag was calculated versus C_L . The results are shown in Figure 1.) below.

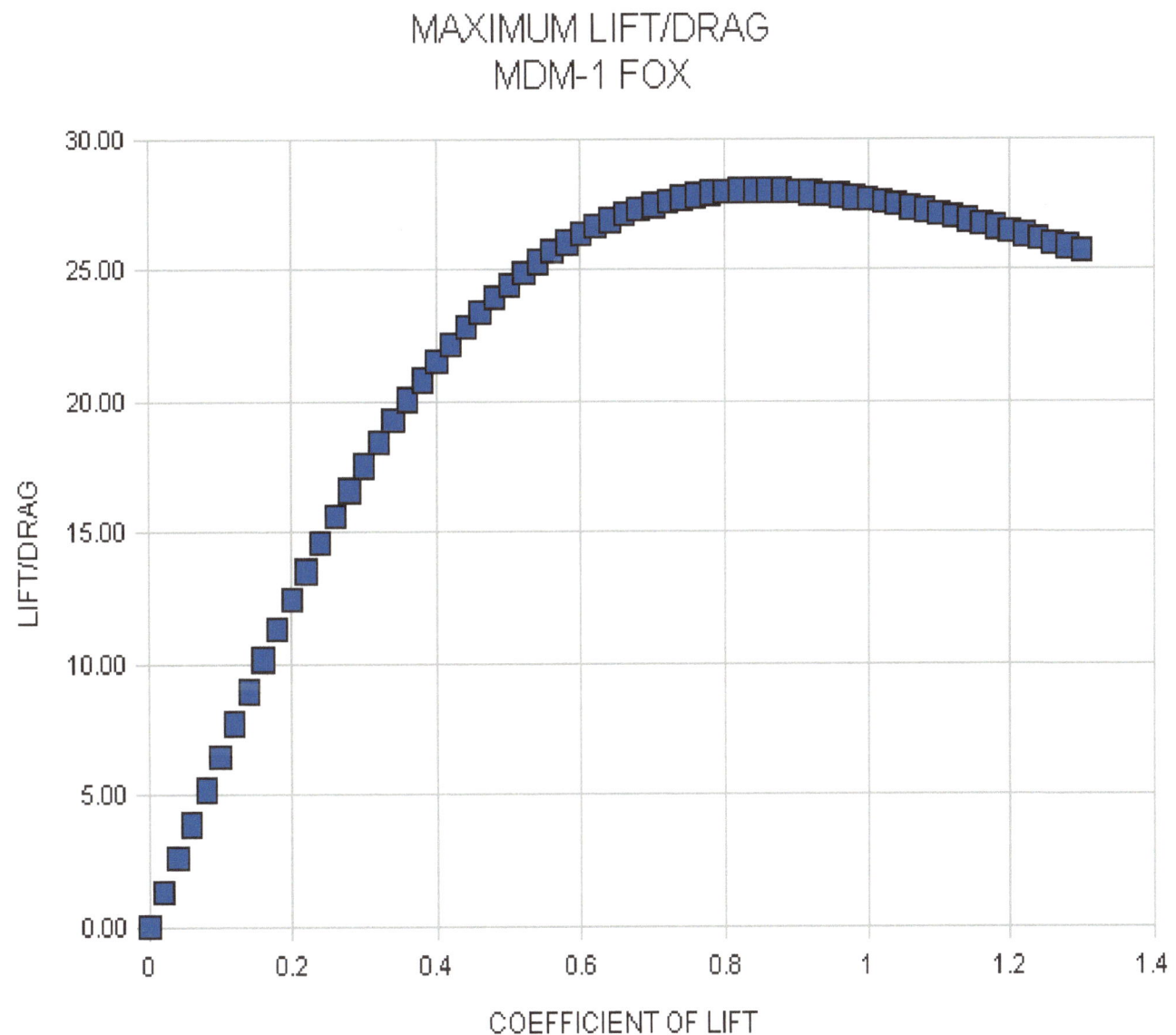

FIGURE 1 LIFT/DRAG VERSUS C_L FOR MDM-1 FOX

Appendix C. Image Analysis of Dynamic-Soaring (DS) Videos

There is a need to maximize load limits on radio-controlled (RC) sailplanes. Aerobatic sailplanes have high load factors compared to sailplanes with high lift/drag ratios used for long distance flights. For example, the aerobatic MDM-1 Fox can take air loads up to 8 g, and the Swift S-1 can take 10 g. But these are piloted aircraft, and need no more load capacity than the limits of the pilot (about 10 g.) So, the question is…if the sailplane is designed to be unmanned, what load limit can be achieved while maintaining reasonably good glider performance? This would involve keeping the aspect ratio of the wing somewhere around 15 or higher, and keeping the wing loading below about 30 Kg/M^2.

DS vehicles are now achieving very high velocities while flying in tight loops. So it would be desirable to determine the g-levels they are successfully experiencing. There are several DS videos on the internet. They typically use radar guns to measure sailplane velocity. An attempt was made to analyze some of these videos for centripetal force. The centripetal acceleration on the vehicle is the square of the velocity divided by the radius of the orbit. The velocity is audibly called out by the operator of the radar gun, and the number of video frames in a complete orbit is counted. The frame rate is also measured, which then yields the period of the orbit. Advancing one frame at a time and measuring the frame rate were done using Microsoft Windows Media Player software. The circumference of the orbit is the velocity times the period. This presumes that the orbit is circular and that the velocity is constant throughout the orbit. These assumptions are seriously in question, but this approach should give a rough indication of centripetal acceleration. Then, the radius of the orbit is the circumference divided by 2π, and the centripetal acceleration can be calculated. These measurements were made on a video (wmv format) obtained at the following link:

http://rcspeeds.com/aircraftdetails.aspx?AC=109

The video was downloaded from the link labeled "**BIG FILE 73mb** Spencer DS's his creation at Weldon." The video shows the Kinetic 100 flown by Spencer Lisenby at Weldon Hill near Lake Isabella, CA. These measurements were made on seven loops. The results are shown below in Table 1. The Windows Media Player displayed the lapsed time from the beginning of the video The video was advanced one frame at a time until the lapsed time changed by one second. This was repeated several times, and the frame rate was consistently measured to be 27 frames per second. The lapsed time from the beginning of the video is shown in the first column of the table. This is the time when the radar operator calls out the velocity (shown in column 2). So it represent the time slightly after completion of the loop. At that time, The video was reversed one frame at a time until the vehicle was at the top of the loop. At that point the vehicle nearly disappeared, because the video shows the rear view of the vehicle heading directly away from the camera. The frame count is begun at that point (in reverse) throughout the entire loop. When the same rear view is obtained, the frame count for that loop is recorded (column 4 of the table.) The frame count is divided by 27 to get the period in seconds, and the calculations described above were used to get the radius of the orbit and the centripetal acceleration. The centripetal acceleration was divided by the acceleration of gravity (32.2 ft/s/s) to obtain the number of gs experienced by the aircraft shown in the last column of the table. The number of digits shown beyond the decimal point is not intended to reflect the accuracy of the measurements.

Table 1. Video Analysis of the Kinetic 100 for Centripetal Acceleration

Time at velocity callout (min:s)	Radar velocity (MPH)	Velocity (ft/s)	Period (frames)	Period (s)	Radius (ft)	Acceleration v^2/r (ft/s/s)	Acceleration (g)
0:04	300	440.0000	85	3.14815	220.4591	878.1675	27.2723
00:13	368	539.7333	67	2.48148	213.1623	1366.6210	42.4416
00:15	361	529.4667	65	2.40741	202.8656	1381.8750	42.9154
00:17	348	510.4000	71	2.62963	213.6119	1219.5400	37.8739
00:20	313	459.0667	71	2.62963	192.1279	1096.8850	34.0647
00:23	315	462.0000	81	3.00000	220.5888	967.6105	30.0500
00:37	362	530.9333	74	2.74074	231.5944	1217.1720	37.8004

The above analysis was repeated with slightly different results. A new frame rate of 30 frames per second was measured and the number of frames per revolution changed also, possibly because the software tended to hang up in moving from one frame to the next. It was necessary to wait up to a minute for the change to occur. The results are shown in Table 2 below. The number of frames per loop is larger in Table 2 than in Table 1. The frame change button is small, so one must be careful not to let the pointer drift off of it. The g-loadings in Table 2 range from 0 to 6% higher than those in Table 1.

Table 2. Second Video Analysis of the Kinetic 100 for Centripetal Acceleration

Time at velocity callout (min:s)	Radar velocity (MPH)	Velocity (ft/s)	Period (frames)	Period (s)	Radius (ft)	Acceleration v^2/r (ft/s/s)	Acceleration (g)
0:04	300	440.0000	89	2.96667	207.7503	931.8882	28.9406
00:13	368	539.7333	72	2.40000	206.1629	1413.0190	43.8826
00:15	361	529.4667	71	2.36667	199.4325	1405.6640	43.6541
00:17	348	510.4000	76	2.53333	205.7895	1265.8960	39.3136
00:20	313	459.0667	77	2.56667	187.5277	1123.7930	34.9004
00:23	315	462.0000	90	3.00000	220.5888	967.6105	30.0500
00:37	362	530.9333	78	2.60000	219.7017	1283.0590	39.8465

The vehicle survived accelerations apparently well in excess of 40 g. Since the vehicle survived, who knows how much further it could go before failure?

This is an amazing structural and aerodynamic feat. Since some DS flights have well exceeded 400 MPH, a similar analysis on those videos was attempted. That effort failed, because the video quality was not as good. They were primarily in the MP4 format (from Vimeo.com) or the flv format (from YouTube.com). Instead of nearly disappearing in one frame at the top of the loop, they disappeared for several frames near the top of the loop. It was not clear whether it was because of the camera being used or the file format.

Appendix D. Model Of Retardation Forces In A Ground Vehicle

The objective is to provide the analytical background necessary for the determination of parameters characteristic of air drag and rolling resistance in a wheeled ground vehicle such as an automobile or truck. These are parameters required in the Project Sea Tree Equilibrium Vector Analysis. Rolling resistance force, F_{rr}, is assumed here to be independent of velocity, and directly proportional to weight and a coefficient of friction or coefficient of rolling resistance, C_{rr}:

1.) $\mathbf{F_{rr}} = -mgC_{rr}$

where m is the mass of the vehicle and g is the acceleration of gravity. Air drag, F_d, has been shown to be:

2.) $\mathbf{F_d} = -0.5\rho\, Av^2 C_d$

where ρ is the air density, $\mathbf{A}$ is the frontal area of the vehicle, $\mathbf{v}$ is the air velocity and $\mathbf{C_d}$ is the coefficient of drag. The minus signs reflect the fact that these are forces of deceleration. Then the total retardation force is $\mathbf{F_{rr}} + \mathbf{F_d}$. Dividing this by the mass yields the acceleration, $\mathbf{dv/dt}$:

3.) $\mathbf{dv/dt} = -0.5\rho\, Av^2 C_d/m - gC_{rr}$

We can now recognize that ρ, $\mathbf{A}$, $\mathbf{m}$ and $\mathbf{g}$ can readily be measured or are known quantities. The relationship between $\mathbf{v}$ and time, $\mathbf{t}$, can be established by determining $\mathbf{C_d}$ and $\mathbf{C_{rr}}$. The $\mathbf{v}$ and $\mathbf{t}$ variables can be separated by dividing by the right side of 3.) and multiplying by $\mathbf{dt}$,

$$\frac{dv}{-(0.5\rho\, Av^2 C_d/m) - gC_{rr}} = dt$$

This is simplified to:

4.) $$\frac{dv}{v^2 + \dfrac{2mgC_{rr}}{\rho \, AC_d}} = -\left(0.5 \, \rho \, Av^2 C_d / m\right) dt$$

The differential equation 4.) is easier to integrate if we introduce the constants:

$$K_1 = \sqrt{\frac{2mgC_{rr}}{\rho \, AC_d}}$$, and

$$K_2 = -0.5 \, \rho \, AC_d / m$$ Then 4.) becomes

5.) $$\frac{dv}{v^2 + K_1^2} = K_2 \, dt$$

Integrating velocity from v_0 to v and time from t_0 to t,

$$\int_{v_0}^{v} \frac{dv}{v^2 + K_1^2} = K_2 \int_{t_0}^{t} dt$$, and this becomes:

6.) $$\arctan(v / K_1) = K_1 K_2 (t - t_0) + \arctan(v_0 / K_1)$$

Assuming estimated values for C_d and C_{rr}, establishes K_1 and K_2. Then, plotting **arctan(v/K$_1$)** versus time should produce a straight line with slope **K$_1$K$_2$** and intercept **arctan(v$_0$/K$_1$)** at **t$_0$**, The definitions of **K$_1$** and **K$_2$** can be rearranged:

7.) $$C_d = -\frac{2mK_2}{\rho \, A}$$

and

8.) $$C_{rr} = -K_1^2 K_2 / g$$

Inserting 7.) and 8.) into 1.) and 2.), **F$_{rr}$** and **F$_d$** become

9.) $\mathbf{F}_{rr} = m \mathbf{K}_1^2 \mathbf{K}_2$

and

10) $\mathbf{F}_d = m \mathbf{K}_2 \mathbf{v}^2$

Considering that the slope obtained from the least-squares fit is $\mathbf{s} = \mathbf{K}_1 \mathbf{K}_2$, (see eq. 6) then equations 9.) and 10.) become

11.) $\mathbf{F}_{rr} = m \mathbf{K}_1 \mathbf{s}$

12.) $\mathbf{F}_d = \dfrac{m\mathbf{s}}{\mathbf{K}_1} \mathbf{v}^2$.

Since mass is presumed to be a known quantity, and the slope is obtained from the least-squares fit, all that is required is an initial estimate of $\mathbf{K}_1$. Then a systematic iterative scheme can be used to minimize the sum of the squares of the deviations from the fit. That will then provide $\mathbf{F}_{rr}$ and the coefficient of $\mathbf{v}^2$ in equation 12.) for $\mathbf{F}_d$.

Appendix E. Data on Retardation Forces in a Ground Vehicle

Considering that the equilibrium vector analysis predicts very large lateral forces on the ground vehicle when towed by a train of four Nimbus 3 sailplanes, a ground vehicle with high lateral traction would be required. A 1992 GMC Sierra truck was available, which has large tires (33-inch diameter and 12.5-inch tread width), and a wide track (72-inches from mid-tread to mid-tread). The wide track relative to the height of the center of gravity provides resistance to roll-over. The vehicle weighs 5700 pounds (25,362 n). The expected lateral force of 101,734 n is over ten times the weight of a Geo Metro, but is only about four times the weight of the GMC Sierra. Resistance to lateral skidding is usually not observed with lateral force above the vehicle weight, so even the GMC Sierra would skid well before maximum power is developed by the Nimbus 3s. A similar problem is faced by racing cars going into a tight turn. They minimize the problem by using air foils with negative angle of attack. The negative lift adds to the vehicle weight and delivers added resistance to lateral forces. The price to be paid includes added aerodynamic drag and a concomitant increase in friction-like rolling resistance, which is directly proportional to weight, but essentially independent of speed. At the high speeds predicted for the tethered sailplane system, aerodynamic drag becomes the dominant retardation force, and rolling resistance is not as important. This aerodynamic concept of increasing resistance to skidding is commonly called "AeroGrip". Its use has been observed to increase skid resistance up to 4 gs (four times the weight of the vehicle).

The analyses of retardation forces in the Geo Metro and the GMC Sierra are being done to provide the characteristics of real vehicles being towed by real sailplanes as an indication of feasibility. The actual design of surface and air vehicles would differ significantly from these vehicles. As for a ground vehicle, it appears that aerodynamic drag could be reduced by reducing the frontal area. The GMC Sierra has a frontal area of about 2.8 m^2. Resistance to lateral skidding could be enhanced by increasing the vehicle weight, increasing the foot-print area of contact between the tires and the ground (although that could depend upon the nature of the ground surface), and the use of an AeroGrip air foil. Most automobiles are designed with a sloping profile that produces a negative coefficient of lift. The resulting traction with the road increases with air speed. That design feature would also be desirable in a towed ground vehicle.

In order to evaluate the performance of the GMC Sierra, and the Geo Metro it was necessary to measure the characteristics involved in retardation forces. The measurement and analysis of those characteristics follows.

Data on a 1992 GMC Sierra

The objective of these measurements and their analysis is to characterize retardation forces in a 1992 GMC Sierra truck. These characteristics are required by the equilibrium vector analysis of a surface vehicle towed by tethered sailplanes. This vehicle is being used in the vector analysis to demonstrate feasibility using existing vehicles. The front and side views of the truck are shown in Figure 1 below.

FIGURE 1 FRONT AND SIDE VIEW OF A 1992 GMC SIERRA

As can be seen, this truck has high clearance and large wheels (16.5-inch rims, 33-inch tire diameter, and 12.5-inch tread width), which makes it a good selection for off-road driving. It weighs about 5,700 pounds (mass = 2588 Kg) when occupied.

DATA COLLECTION

The approach used in this experiment was to bring the truck up to speed (as high as road conditions will allow), put it into neutral, and measure velocity and time as it coasts down to a stop. This was done on a flat, level surface to eliminate the accelerating or decelerating effects of gravity. Then the model for retardation forces in Appendix D was used to obtain the pertinent characteristics for this truck.

Dry lake beds are level over considerable distances, so the Lucerne Dry Lake near Lucerne Valley, CA was chosen for the location of the experiment. This dry lake is in the extreme southern portion of the Mojave Desert in southern California. Velocity was measured using the speedometer in the dash board of the truck. The indicated velocity was recorded using a Sony HANDYCAM model DCR-DVD108. This video camera has a time-and-date stamp that was used to measure time in integral seconds. The camera records at the rate of 30 frames per second. Video display software (Windows Media Player by Microsoft) on a personal computer was used to count the frames following the change in the displayed seconds of time. This provided the time with a precision of one thirtieth of a second.

The experiment was conducted late in August of 2010. The high temperature for the day was predicted to be 109 °F in Lucerne Valley, so it was conducted early in the morning at a temperature in the high 80s. The truck was brought up to a speed of about 40 MPH. Above that speed, irregularities in the surface caused excessive jostling of the camera. The left hand of the operator was on the steering wheel, and the right hand held the camera. The problem was to move the gear shift into neutral. It was done with the hand holding the camera. Then the camera was pointed at the speedometer. Although the lake bed was smooth enough to travel at 40 MPH, the ride was somewhat bumpy, and the hand-held camera did not provide a steady view of the speedometer. However that was not necessary, because the video was inspected one frame at a time. All that was required was a steady view during a few frames out of every fifty to one hundred frames. An example is shown in Figure 2 below of a frame taken during one of the runs. This method was sufficient to acquire twenty three to thirty nine speed measurements per run. It was repeated three times, providing a total of three runs. Note from Figure 2 that the speedometer scale is marked with finest divisions of 1 MPH except below 10 MPH, where the finest divisions are 2 MPH. Data below 10 MPH were regarded as unreliable, and were discarded. As the velocity approached zero, the camera was pointed out the window, toward the soil at a distance of about 10 feet. At very low velocities, the camera was very steady, and the instant when the truck came to a stop was readily measured.

Figure 2. Sample Video Frame

Although the lake bed was flat, there were a few gullies about a foot wide, every few hundred feet. Impact with them jolted the truck severely. One of the lessons learned was to make a slow, dry run over the path the truck will travel. A path without gullies would be selected for a run at higher speed. The velocity and time data are shown in Table 1 below.

Table 1 VELOCITY VERSUS TIME DATA

run #1				run #2				run #3			
velocity (MPH)	camera display time (s)	frames past display time (s/30)	time (s)	velocity (MPH)	camera display time (s)	frames past display time (s/30)	time (s)	velocity (MPH)	camera display time (s)	frames past display time (s/30)	time (s)
34	57	24	57.800	39	47	26	47.867	36	27	26	27.867
32.8	60	10	60.333	38.5	48	19	48.633	34.7	30	2	30.067
32	61	13	61.433	38	49	17	49.567	33.3	32	4	32.133
31.5	62	18	62.600	37.5	50	4	50.133	32.4	33	12	33.400
30.8	64	2	64.067	37	50	29	50.967	31.4	34	28	34.933
30.3	64	20	64.667	36	52	12	52.400	29.3	38	28	38.933
30	64	26	64.867	35	54	1	54.033	28.5	40	12	40.400
29	66	26	66.867	34	55	16	55.533	27.7	42	7	42.233
27.8	68	22	68.733	33.1	57	4	57.133	27.5	42	22	42.733
27.9	69	5	69.167	32.5	58	11	58.367	26.7	44	8	44.267
27.6	69	17	69.567	32	59	7	59.233	26.4	45	9	45.300
27	71	5	71.167	31	60	11	60.367	25.2	47	19	47.633
26.5	72	14	72.467	29.9	62	26	62.867	24.1	50	2	50.067
24.7	76	14	76.467	29	64	29	64.967	23.4	52	4	52.133
24.1	77	19	77.633	28.5	65	14	65.467	21.7	55	10	55.333
24	78	0	78.000	27.6	67	18	67.600	21	57	9	57.300
23.5	79	5	79.167	27	69	4	69.133	20.1	59	29	59.967
22.5	81	18	81.600	26.5	70	8	70.267	18.9	63	0	63.000
22	82	28	82.933	26	71	12	71.400	17.7	65	18	65.600
21.5	84	8	84.267	24.9	73	26	73.867	15.1	73	1	73.033
21	85	19	85.633	23.8	76	21	76.700	14.2	75	26	75.867
18.9	90	11	90.367	23	78	22	78.733	13.4	78	6	78.200
18.4	91	10	91.333	22.5	80	10	80.333	0	121	0	121.000
18	93	0	93.000	22	81	13	81.433	N₃ = 23			
17.5	94	8	94.267	21.6	82	19	82.633				
17	95	21	95.700	21.1	84	1	84.033				
15	101	12	101.400	19.8	87	7	87.233				
14	104	18	104.600	19.1	88	15	88.500				
13	107	20	107.667	18.5	90	3	90.100				
12	110	28	110.933	18.1	91	17	91.567				
11	114	2	114.067	17.6	93	11	93.367				
0	147	15	147.500	17	94	22	94.733				
N₁ = 32				16.3	96	25	96.833				
				16	98	11	98.367				
				15	101	8	101.267				
				14	104	4	104.133				
				13.5	105	17	105.567				
				13.1	107	1	107.033				
				0	146	1	146.033				
				N₂ = 39							

DATA REDUCTION

The model for retardation forces in Appendix D, indicates that the relationship between velocity and time can be made linear by taking $ATAN(v/K_1)$ versus time, where

$$1.)\ K_1 = \sqrt{\frac{2mgC_{rr}}{\rho A C_d}}$$

The term m is the mass of the vehicle, g is the acceleration of gravity, ρ is the air density, A is the frontal area of the vehicle, v is the air velocity, C_{rr} is a coefficient of friction or a coefficient of rolling resistance, and C_d is the coefficient of drag. In order to use the linear relationship between ATAN(v/ K_1) and time, it is necessary to make a first estimate of K_1. The terms of equation 1 are known, measured or estimated. The mass of the truck is known to be 2588 Kg, g is known to be 9.8 m/s^2, ρ is 1.2 Kg/m^3 at sea level. Lucerne Dry Lake is at an elevation of 3,000 ft. so $\rho = 1.1$ Kg/ m^3. The frontal area, A, was measured to be about 2.8 m^2. The coefficients were estimated from literature data to be $C_{rr} = 0.01$, and $C_d = 0.6$. This yields the first estimate $K_1 = 16.6$ m/s. This estimate need not be accurate, because an iterative scheme will be used to converge upon a value of K_1 that results in the minimum sum of the squares of deviations from the line fit to the data. So K_1 will be an output of the data analysis. This estimate of K_1 was used to calculate ATAN(v/ K_1) with the data on velocity shown in Table 1. The results are plotted in Figure 3.

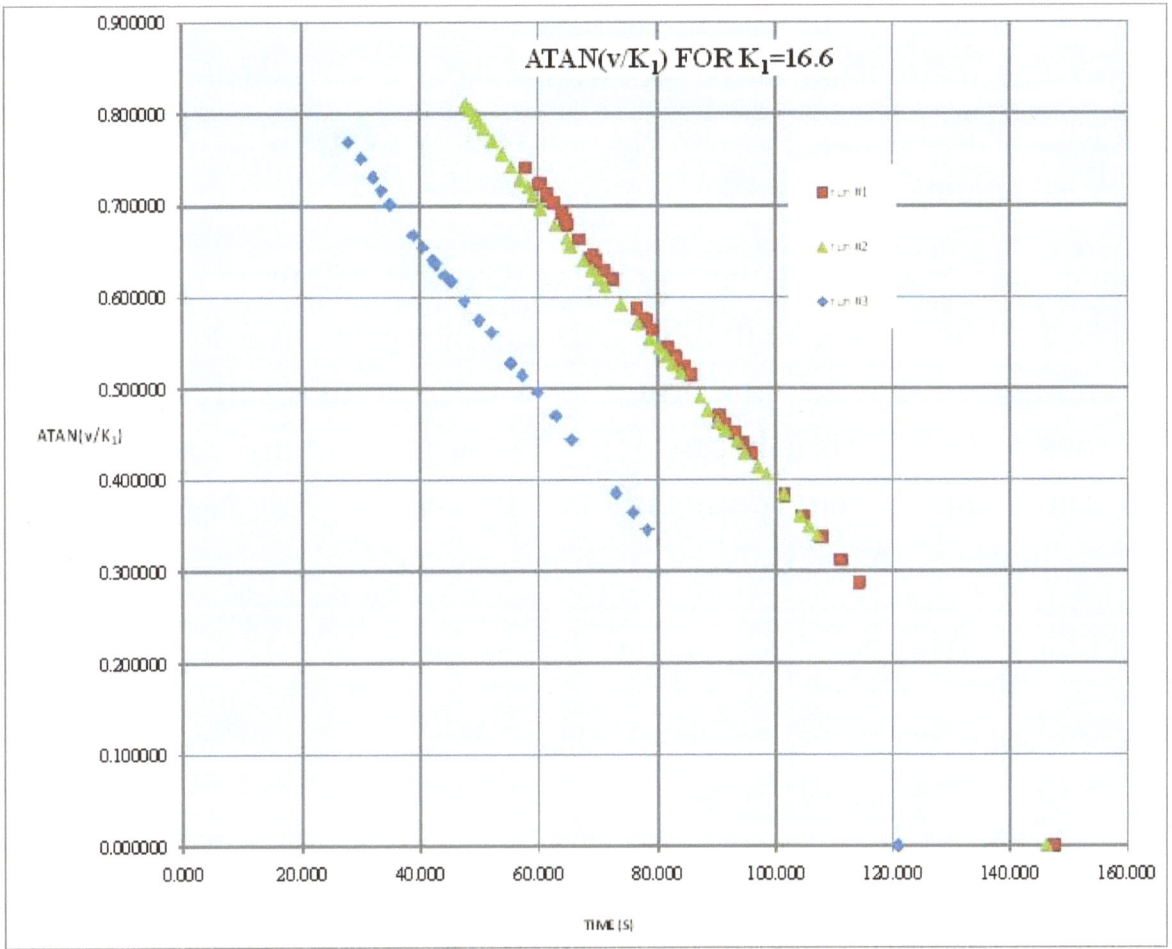

FIGURE 3 tan-1(v/K1) VERSUS TIME FOR ALL THREE RUNS

The model assumes that the retardation forces include only a rolling resistance, F_{rr}, which is independent of velocity, and aerodynamic drag, F_d which is proportional to the square of velocity. Defining s to be the slope of the linear relationship between $ATAN(v/K_1)$ and time the model concluded that :

2.) $F_{rr} = mK_1s$

, and

3.) $F_d = \dfrac{ms}{K_1}v^2$

Since mass is presumed to be a known quantity, and the slope is obtained from the least-squares fit, all that is required is an initial estimate of K_1. Then a systematic iterative scheme can be used to minimize the sum of the squares of the deviations from the fit. That will then provide F_{rr} and the coefficient of v^2 in equation 3 for F_d. These are the parameters required by the equilibrium vector analysis of a surface vehicle towed by tethered sailplanes.

Before the least-squares fit is obtained, the time data from all three runs must be translated to a unified scale, so that the data can be treated as a single body. It is clear from Figure 3 that the time origin for run 3 is significantly different than that for runs 1 and 2. This is because the time origin is entirely arbitrary for each run. The number of data points in run 1 is N_1 ($N_1 = 32$), N_2 in run 2 ($N_2 = 39$) and N_3 in run 3 ($N_3 = 23$). Now define y_1, y_2 and y_3:

$$y_{1,i} = \tan^{-1}\left({v_{1,i}}/{K_1} \right) \text{ for } i = 1 \text{ to } N_1$$

$$y_{2,i} = \tan^{-1}\left({v_{2,i}}/{K_1} \right) \text{ for } i = 1 \text{ to } N_2$$

$$y_{3,i} = \tan^{-1}\left({v_{3,i}}/{K_1} \right) \text{ for } i = 1 \text{ to } N_3$$

where $v_{1,i}$ is the ith observation of velocity in run 1...etc. Similarly $t_{1,i}$ is the ith observation of time in run 1...etc. Define T_1, T_2 and T_3 to be translations of the time axis for runs 1, 2 and 3 respectively. After finding the best translations, the translated data will be fit to a line of the form y=a+st, where a is a constant determined by the least-squares fit, along with the slope, s, defined above. The time deviation from the fit for each $t_{1,i}$ is

$$t_{1,i} - \frac{y_{1,i} - a}{s}$$

Now choose T_1 to be equal to the value of this deviation averaged over the data of run 1:

4.) $T_1 = \langle t_1 \rangle - \dfrac{\langle y_1 \rangle - a}{s}$

Similarly,

5.) $T_2 = \langle t_2 \rangle - \dfrac{\langle y_2 \rangle - a}{s}$

6.) $T_3 = \langle t_3 \rangle - \dfrac{\langle y_3 \rangle - a}{s}$

The angular brackets in these equations represent the value of the contained quantities averaged over the observed data, e.g.

$$\langle t_1 \rangle = \frac{1}{N_1} \sum_{i=1}^{N_1} t_{1,i}$$

Translation of the three time axes in accordance with equations 4, 5 and 6 will result in a zero value for the average time deviation from the fit

Equations 4, 5 and 6 provide time-axis translations in terms of average values of t and y, and in terms of the parameters of the least-squares fit to the data. However, a reasonably good fit cannot be made until reasonably good translations are applied to the time data. After obtaining the first fit parameters, a new set of translations can be obtained from equations 4, 5 and 6. This will lead to an iterative relationship between the fit parameters and the set of translations. Each new set of translations will lead to new fit parameters, and each set of fit parameters will lead to a new set of translations. Presumably this process will converge upon parameters that no longer change from one iteration to the next. To begin this process, a first estimate of the translations was made as follows:

$T_1 = <t_1> - <t> = 10.054$

$T_2 = <t_2> - <t> = 3.6026$

$T_3 = <t_3> - <t> = -20.097$

where the average of time measurements for all three runs is $<t>$.

Using these first estimates for time translation, and the first estimate for $K_1 = 16.6$ m/s, (see that estimate early in the section on data reduction), the following iterative sequence (Table 2) was calculated:

Table 2 ITERATION ON TIME TRANSLATION

iteration	K_1	T_1	T_2	T_3	slope	RMS error
0	16.6	10.0541667	3.6025641	-20.0971	-0.008126	0.0180709
1	16.6	7.3566642	5.8565322	-20.1660	-0.008126	0.0048698
2	16.6	7.3566642	5.8565322	-20.1660	-0.008126	0.0048698

The RMS error column represents the root-mean-squared deviation in y from the fit. This table indicates that, while holding K_1 constant at 16.6 m/s, calculation of the least-squares fit produces time translations that converge quickly. Now, change K_1 until a minimum RMS error is achieved. This is done using a tool called "solver" in the Microsoft Excel spreadsheet. This changed K_1 systematically while holding time translations constant, to minimize RMS error. Care must be taken to setup solver to allow a sufficient number of changes in K_1, to provide sufficient precision in the calculations, and to require a sufficiently narrow convergence criterion. Otherwise, a premature convergence can lead to considerable error. These calculations permitted 300 iterations, a precision of 10^{-8} and a convergence in K_1 of 10^{-6}. That process was considered to be one iteration. It yielded linear fit parameters for that set of time translations, and also yielded a new set of translations. The new translations were used in the next iteration, again using the solver tool. These results are shown in Table 3 below.

Table 3 ITERATION ON K1

iteration	K_1	T_1	T_2	T_3	slope	RMS error
1	16.882938	7.356664	5.856532	-20.166001	-0.008050	0.00484129
2	16.887102	7.347015	5.864893	-20.166753	-0.008049	0.00484087
3	16.887209	7.346767	5.865105	-20.166767	-0.008049	0.00484087
4	16.887209	7.346761	5.865110	-20.166767	-0.008049	0.00484087
next		7.346761	5.865110	-20.166767		

The data with translated time axis are plotted along with the least-squares fit in Figure 4 below. This yields the desired parameters for the 1992 GMC Sierra, K_1=16.89 and s=-0.008049. Since the mass of the vehicle is known to be 2588 Kg, equations 2 and 3 become

$$F_{rr} = -352 \ n$$

$$F_d = -1.233 \ v^2 \ n$$

where v is in units of m/s.

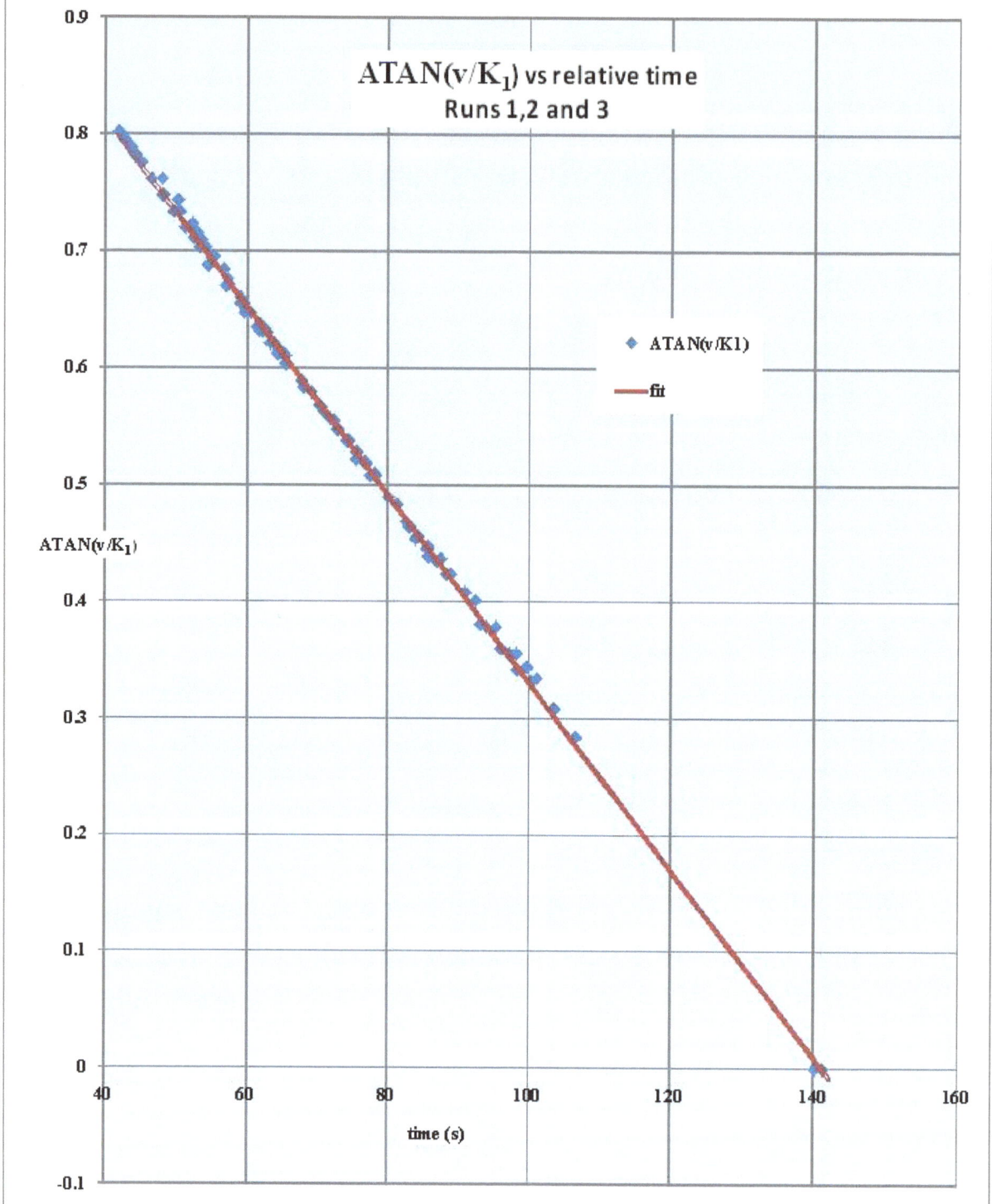

FIGURE 4 GMC SIERRA tan-1(v/K1) VERSUS TIME WITH FIT

Data on a 1992 Geo Metro

The Geo Metro is a small car which should offer much lower retardation force than the GMC Sierra. This should occur, because the former has lower frontal area, leading to reduced aerodynamic drag, and lower weight, leading to lower rolling resistance. The determination of the retardation characteristics of this vehicle will provide a first step in evaluating the influence of pertinent parameters of the ground vehicle in a tethered sailplane system. In addition, it is expected to further demonstrate feasibility of such systems.

DATA COLLECTION

Data on retardation forces in a 1992 Geo Metro were shown on the internet at http://www.iwilltry.org/b/how-to-measure-the-drag-coefficient-of-your-car/ published by Rob Steves on August 30, 2007. He brought the car up to speed, put the gear shift in neutral, and used a stopwatch to measure elapsed time as the velocity of the car decayed. He took velocity measurements at 10 second intervals, beginning at 70 KPH. He took a sequence of data on a level road going in one direction, then repeated the sequence going in the opposite direction. These data are shown in Table 4 below.

Table 4 Geo Metro Velocity Versus Time Data

time (s)	Velocity (KPH)					
	run #1	run #2	run #3	run #4	run #5	run #6
0	70	70	70	70	70	70
10	61	60	60	60	61	60
20	52	52	51	51	52	51
30	44	44	43	43	43	44
40	37	37	38	37	37	37.5
50		32	32		32.5	32
60		27			27.5	27
70		22			22.5	

DATA REDUCTION

As in the previous section on the GMC Sierra, a linear relationship with time is obtained by using $\tan^{-1}(v/K_1)$ as the dependent variable, where K_1 is given by equation 1 in that section. To make that change of variable, it is necessary to make an estimate of K_1. The mass of the vehicle is m=1000 Kg. The acceleration of gravity is 9.81 m/s^2 . The air density is 1.22 kg/m^3. The frontal area of the vehicle is A=2.3 m^2, and the retardation coefficients are estimated to be C_{rr}=0.01 and C_d=0.4. These values lead to a first estimate of K_1=13.22. (See equation 1.) Using this estimate, $\tan^{-1}(v/K_1)$ was calculated for each velocity value in Table 4.

At this point, the analysis of the GMC Sierra data in the previous section translated the very different time axes for three velocity-versus-time runs. An effort was made in collecting data on the Geo Metro to start the time axis immediately at a velocity of 70 KPH for each of six runs. Therefore, the first data point was always v=70 KPH and t=0 seconds for each run. Presumably, that would eliminate the need for a time translation. However, there is no reason to believe that the experimental error in the first data point is any different than that for any other point. Therefore, this analysis of the data includes six time translations.

Calculation of the translations, using $y = \tan^{-1}(v/K_1)$, were done in accordance with equations 4, 5 and 6 in the previous section, supplemented by similar equations 7, 8 and 9 for runs 4, 5 and 6:

$$7.) \ T_4 = \langle t_4 \rangle - \frac{\langle y_4 \rangle - a}{s}$$

$$8.) \ T_5 = \langle t_5 \rangle - \frac{\langle y_5 \rangle - a}{s}$$

$$9.) \ T_6 = \langle t_6 \rangle - \frac{\langle y_6 \rangle - a}{s}$$

Once again, these translations are in terms of the parameters of the least-squares fit, a and s, and the fit is dependent on the translations. Therefore, there is an iterative relationship between the fit and the translations requiring a first estimate for each. The estimate for $K_1=13.22$ (see above) is all that is required to get the fit. Since an effort was made to start the time axes at zero, the first estimate of the translations were all taken to be zero. Holding K_1 constant, the linear fit was obtained. The parameters of the fit were used in equations 4 through 9 to obtain the next guess for the time translations. Using this simple iteration (i.e. next guess = previous answer), the iterative process converged rapidly. The translations converged to well within one millisecond, and the slope converged to well within 0.1% in three iterations as shown in Table 5.

Once again, the RMS error column represents the root-mean-squared deviation in y from the fit.

Table 5. Geo Metro Iteration on Time Translations

Iteration	K_1 (m/s)	Translation (s)						Slope (s^{-1})	RMS error
		T_1	T_2	T_3	T_4	T_5	T_6		
0	13.22	0	0	0	0	0	0	-0.0076871	0.0054678
1	13.22	0.2343556	-0.0218162	-0.1419734	-0.5243009	0.3085700	0.0010766	-0.0077012	0.0050961
2	13.22	0.2178640	-0.0102034	-0.1485609	-0.5393949	0.3195741	0.0034364	-0.0077025	0.0050953
3	13.22	0.2164285	-0.0091926	-0.1491343	-0.5407088	0.3205319	0.0036418	-0.0077026	0.0050953
next		0.2163038	-0.0091048	-0.1491841	-0.5408229	0.3206151	0.0036596		

Table 5 provides time translations for the case K_1=13.22. Now, it is necessary to iterate on K_1 to minimize the RMS error. This is done while holding the translations constant and systematically changing K_1 until minimum RMS error is achieved using the solver tool in the Excel spreadsheet. This constitutes one iteration. Then the new translations corresponding to the newly calculated fit parameters are used for the next iteration. The results are shown in Table 6. Using the newly calculated translations as the guess for the next iteration is simple iteration which converged more slowly than in Table 5. So an accelerated convergence scheme was used. It was assumed that the answer was linearly dependent upon the guess, and that relationship was projected to an intersection with the line of equality, i.e. the line where the answer is equal to the guess. This projected guess was used after four simple iterations, whereupon convergence was instantly achieved.

Table 6 Geo Metro Iteration on K1

Iteration	K_1 (m/s)	Translation (s)						Slope (s^{-1})	RMS error
		T_1	T_2	T_3	T_4	T_5	T_6		
1	12.9534800	0.216304	-0.009105	-0.149184	-0.540823	0.320615	0.003660	-0.0077168	0.005073
2	12.9397262	0.228837	-0.026605	-0.133909	-0.525890	0.305550	0.008166	-0.0077160	0.005072
3	12.9379874	0.231151	-0.028692	-0.132444	-0.523590	0.303653	0.008168	-0.0077159	0.005072
4	12.9377539	0.231466	-0.028972	-0.132249	-0.523278	0.303398	0.008165	-0.0077159	0.005072
project		0.231508	-0.029009	-0.132223	-0.523236	0.303363	0.008164		
5	12.9377175	0.231515	-0.029015	-0.132219	-0.523230	0.303358	0.008164	-0.0077159	0.005072
project		0.231515	-0.029015	-0.132219	-0.523230	0.303358	0.008164		
6	12.9377175	0.231515	-0.029015	-0.132219	-0.523230	0.303358	0.008164	-0.0077159	0.005072
next		0.231515	-0.029015	-0.132219	-0.523230	0.303358	0.008164		

Although an attempt was made to coordinate the time axes from run to run, the data indicate that translations ranging up to about a half second are required. The data with translated time axes are plotted along with the least-squares fit in Figure 5 below. This yields the desired parameters for the 1992 Geo Metro, K_1=12.938 and s=-0.007716. Since the mass of the vehicle is known to be 1000 Kg, equations 2 and 3 become

$$F_{rr} = -99.8 \text{ n}$$

$$F_d = -0.596v^2 \text{ n}$$

where v is in m/s. The excellent linear fits in Figures 4 and 5 confirm the assumption that viscose forces are not significant in these ground vehicles. At least, that appears to be true below about 20 m/s. That still needs to be confirmed at higher velocities.

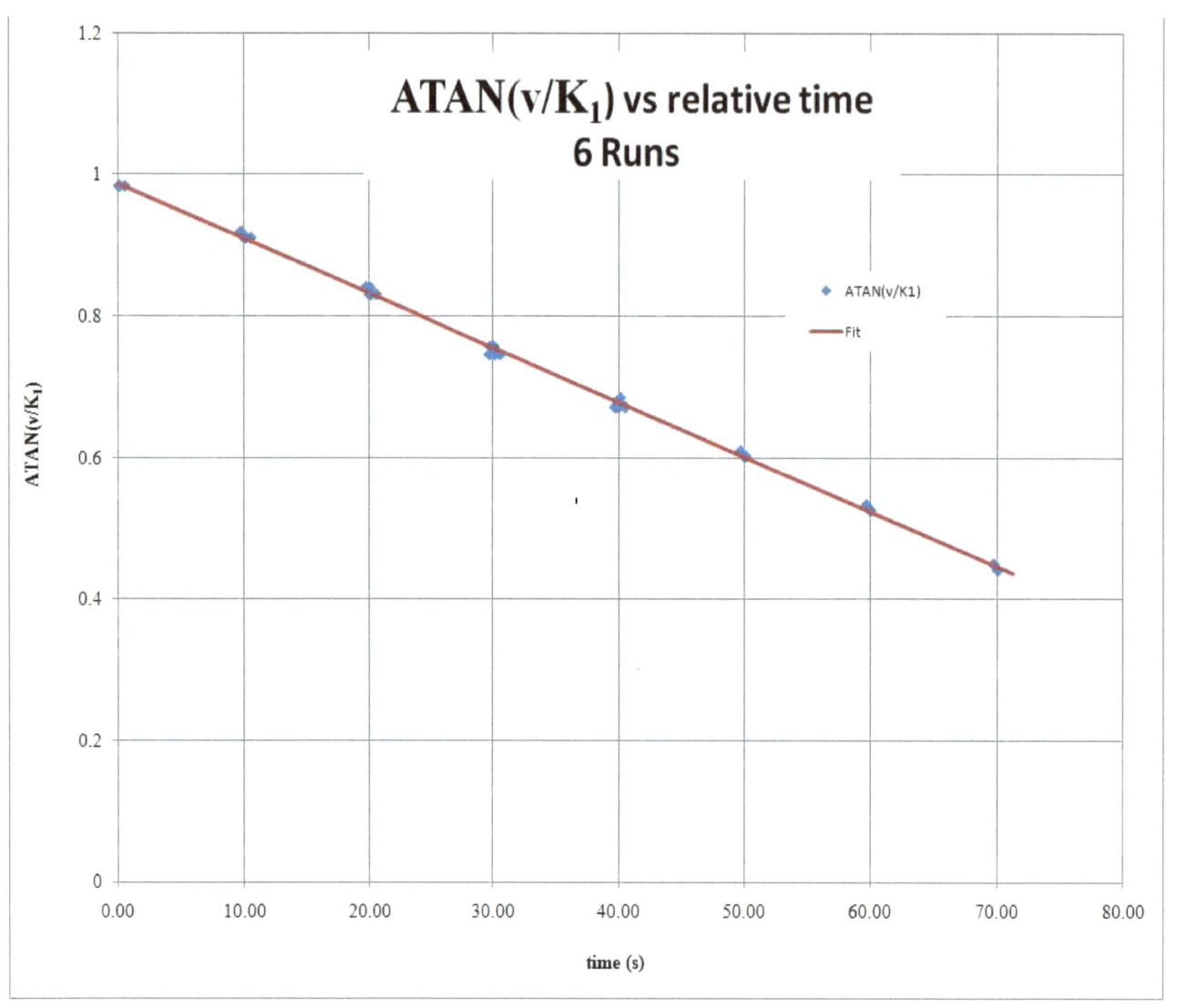

FIGURE 5 GEO METRO tan-1(v/K1) VERSUS TIME WITH FIT

Appendix F. MRK Air-Vehicle Stability

Obviously, air-vehicle stability is essential to the viability of the MRK concept. Instability in pitch, roll and yaw must be controlled. Careful design which places the tether attachment forward of the center of gravity and the center of lift should be adequate for pitch stability. Attaching the tether to the air vehicle through a bridle helps to provide roll stability. The bridle also distributes the tether tension to two, separated positions on the wing rather than one position in the absence of a bridle. That feature increases the load limit of the air vehicle, which is an important factor in the performance of the MRK system. In addition, mechanisms used to control roll can be used in conjunction with rate gyros to achieve roll stability. Control of roll can be achieved using ailerons or an adjustable bridle. The concept of an adjustable bridle is shown in Figure 1, below.

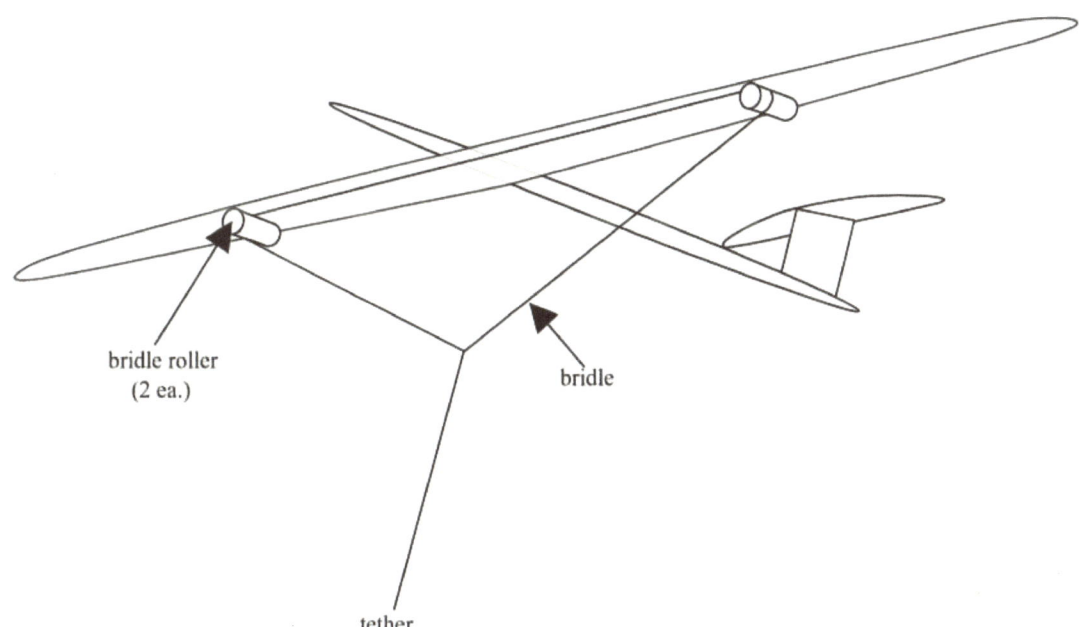

bridle roller
(2 ea.)

bridle

tether

FIGURE 1 ADJUSTABLE BRIDLE LAYOUT

The motor driven rollers cause the vehicle to roll by moving the bridle-tether attachment left or right. The rollers could be recessed into the body of the wing to reduce drag. The bridle could be terminated at each roller rather than forming a complete loop around both rollers. If ailerons are used for roll control, the use of a bridle may result in large resistive moments. The distance between the tether attachment point and the center of lift forms the corresponding moment arm resulting in a moment that must be overcome to change roll. The MRK system is expected to produce very large tether tension in reaction to very large air loads. Consequently, the moment is expected to be very large. The moment arm and the corresponding moment are much smaller when roll is controlled by an adjustable bridle.

A more daunting task is to achieve yaw stability. Three approaches to yaw stabilization are envisioned for MRK air vehicles. Firstly, a rate gyro is used to produce a signal proportional to the rate at which the heading of the air vehicle is changing. This signal is then used to change the angular position of the rudder, thereby holding the heading nearly constant. Secondly, the technique of sheeting (discussed in section 4 above) has the effect of producing stability. A departure in the heading due to undesired yaw in a tethered air vehicle can result in a catastrophic dive into the ground. But, controlled changes in heading can result in a sequence of controlled dives and climbs (or just loops) that obscure those yaw instabilities. The apparent air speed of an MRK air vehicle at the end of a controlled dive would be much higher than the average air speed. Since power generation goes up with the cube of the air speed, an advantage in average power generation is realized. The air loads on the MRK air vehicle would also be much higher than the average at the end of a controlled dive. Since many power generation scenarios are load limited, sheeting may not be advantageous for those particular MRK systems.

Thirdly, a technique is presented here for using gravity as a reference to control heading. The concept, referred to as the Pendulum Nose Rudder (PNR), uses a rigid, physical pendulum to position a nose rudder. This is in contrast to a simple pendulum with a weight attached to the end of a flexible cable. Say for example, that an undesired yaw in the air vehicle occurs to port. A rudder attached to a pendulum would be held vertical, as nearly as possible (presuming that the rudder is held parallel to the pendulum and that the rotation axis of the pendulum fulcrum is not parallel to gravity). Such a rudder would experience an aerodynamic force to starboard, created by the motion of the vehicle to port. But, a pendulum-rudder mounted on the tail structure, would push the tail to starboard, resulting in a heading even further to port. This is an unstable situation. However, if we mount the pendulum-rudder on the nose, a correction to the heading occurs to starboard, and the tendency is toward yaw stability. Obviously, the same argument applies for an undesired yaw to starboard.

An attempt was made to demonstrate the PNR concept using a light-weight, diamond-shaped kite (see Figures 2 and 3). The frame was constructed from pine, with a 1.524 M spar and a 1.753 M spine. The spar was straight, without bow. The rip-stop cover (mass per unit area of 0.043 Kg/M^2) was bonded to peripheral string to form the body of the kite. The physical pendulum and the nose rudder were mounted on the dorsal side of the kite. The pendulum extended 0.762 M aft of the pivot. A longer pendulum was tried, but when undesired yaw occurred to one side, the pendulum swung to the same side, resulting in a shift of the center of gravity to that side, which in turn, caused roll to that side. That roll was detrimental to stability. Future versions will require a horizontal stabilizer attached to the end of the pendulum. That will cause the center of lift to swing to the same side as the center of gravity. A small vertical stabilizer was mounted ventrally at the tail end. However, the more massive dorsal rudder and pendulum raised the center of gravity slightly dorsal relative to the center of lift. This is undesirable for stability, and future versions should use a ventral mounting of the rudder and/or the pendulum. The design will then have to avoid interference between the pendulum and the tether (e.g. use a bridle or place the tether at the pivot.) The center of gravity was 0.0032 M aft of the spar, and the center of lift was 0.0238 M aft of the spar. The tether was attached right at the intersection of the spar and the spine without a bridle. The kite area was 1.335 M^2, and the mass was 0.407 Kg, corresponding to a wing loading of 0.305 Kg/M^2. The wing loading was roughly 50% higher than that desired of a typical, stable kite.

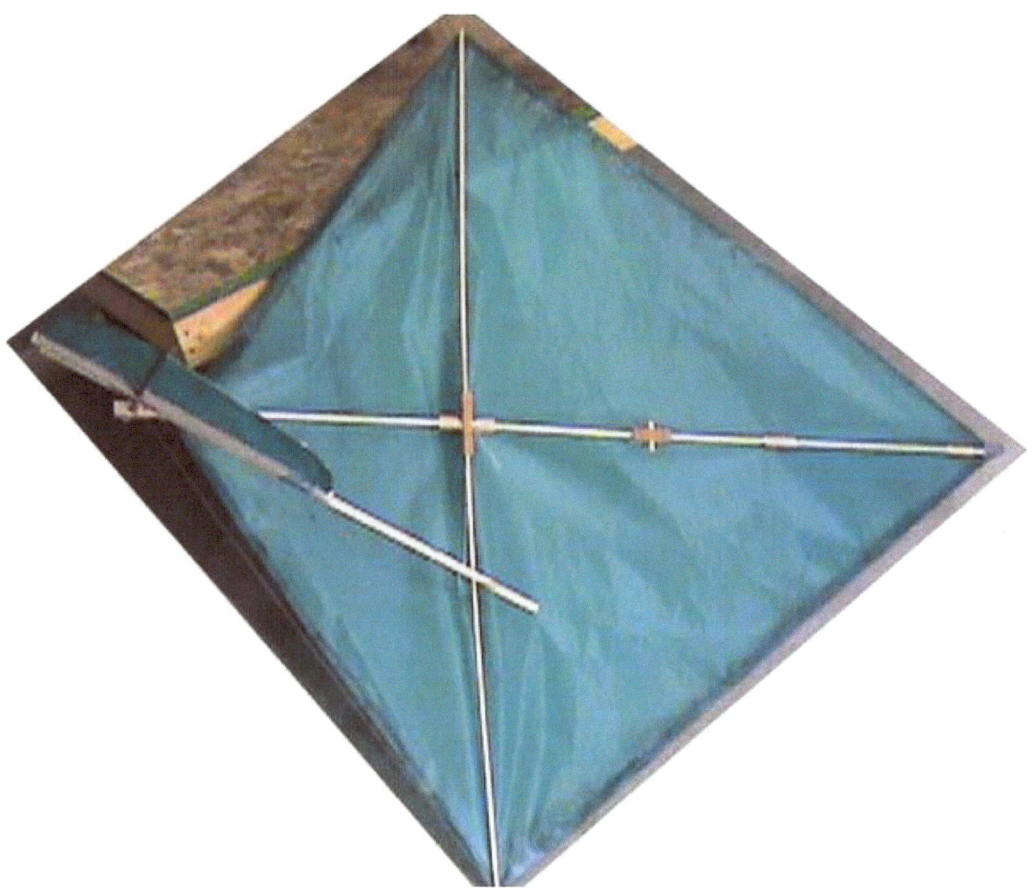

FIGURE 2. DORSAL SIDE OF PNR DIAMOND KITE

FIGURE 3. VENTRAL SIDE OF PNR DIAMOND KITE

The kite was flown in gusty wind. The wind speed was not measured, but it was probably around 3 M/S, with considerable variability. The flight started out very well, with the kite rising to a height of 12 to 15 M. The PNR worked well in stabilizing the kite. After about 5 minutes of stable flight, the wind dropped off, and the kite fell into a nose-first dive. It crashed into the ground catastrophically, fracturing the spine in two places. However, stabilization of the kite using the PNR concept was demonstrated. It is notable that stability was achieved without a bridle, without bow in the spar, and without a tail, under the adverse conditions of a dorsal center of gravity, a high wing loading and a shifting center of gravity (caused by the pendulum) tending to cause destabilizing roll. One feature of the diamond design that lends some tendency toward stability is the very low aspect ratio (1.79).

A new PNR design was undertaken to correct the dorsal center of gravity and the high wing loading of the diamond kite, and to demonstrate stability in a kite with a higher aspect ratio, closer to that needed in the MRK system. A delta kite was chosen with a spar at the trailing edge, thus moving the center of gravity aft. The schematics are shown in Figures 4 and 5. The wing span is 1.8288 M, the spine length is 0.6096 M, the rip-stop covered wing (same cloth as for the diamond) has an area of $0.5574 M^2$, the distance forward of the spar to the rudder/pendulum pivot is 0.4572 M, to the center of gravity is 0.178 M, and to the center of lift is 0.203 M. The dorsal nose rudder and the dorsal tail fin are both semicircular with a diameter of 0.3048 M. Both the spar and the spine are Carbon Fiber Reinforced Plastic (CFRP) to take advantage of the high strength-to-weight ratio of that material. The whole-vehicle mass is 0.0855 Kg and the wing loading is $0.1534 Kg/M^2$ which should help in achieving stability. The aspect ratio is 6, well above that for the diamond kite. The physical pendulum is mounted well below the wing in a ventral position to keep the center of gravity ventral to the wing. The pendulum is shown aligned parallel to the rudder, and the pendulum extends 0.762 M aft of the pivot. The tether is attached at the pivot, well forward of the centers of lift and gravity.

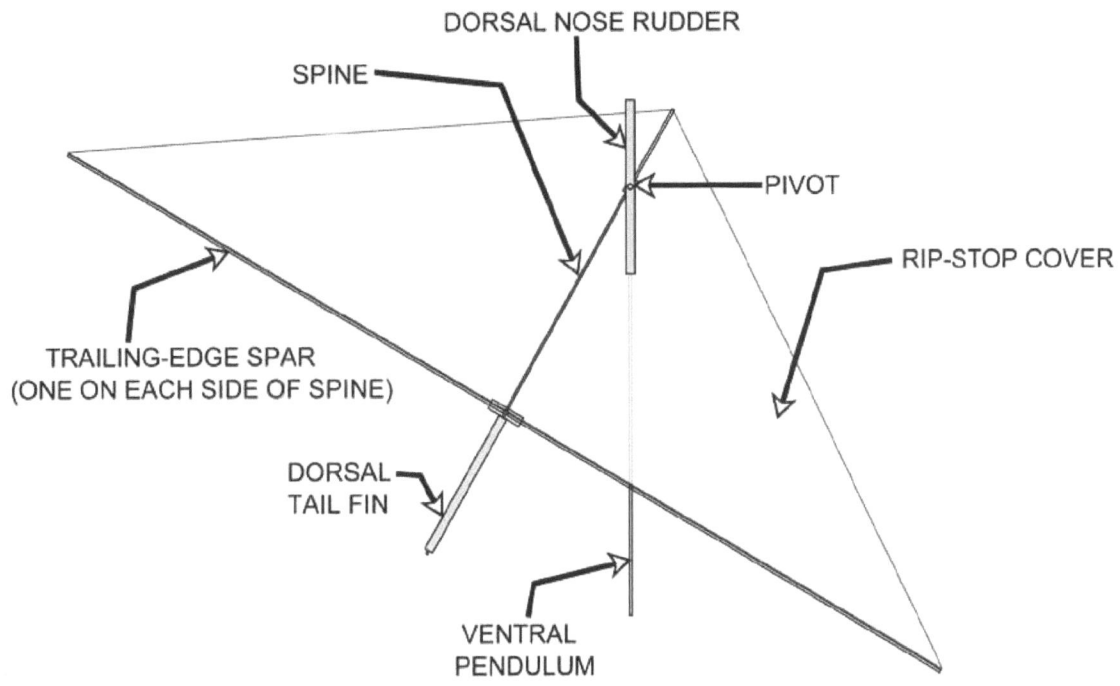

FIGURE 4. DORSAL SIDE OF PNR DELTA KITE

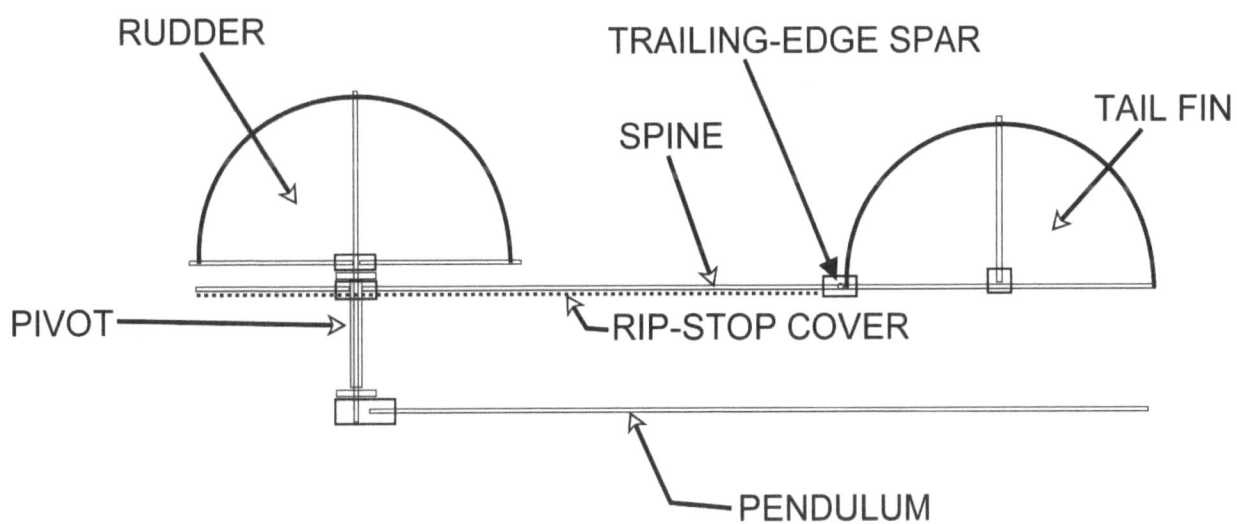

FIGURE 5. SIDE VIEW OF PNR DELTA KITE

This design still has some problems. Figure 4 shows undesirable yaw to starboard. The corresponding motion of the kite to starboard will produce an aerodynamic force to port on the sides of the rudder and the tail fin. This force is intended to correct the heading. However it also produces a moment about their longitudinal axes tending to roll the kite to port. This is constructive, helping to produce stability. But the dorsal mounting of these fins may reduce their effectiveness in cases where the angle of attack is large (typical of these kites) resulting in turbulent, detached flow over the dorsal side of the wing. When the angle of attack is small, flow is not detached, but the axis of the pivot becomes nearly vertical, where gravity has very little effect on the pendulum, if any. These fins may be more effective mounted in a ventral position. Another problem is that when the pendulum shifts away from the neutral position, there is a concomitant, destabilizing lateral shift in the center of gravity. In the above discussion of the diamond kite, it was suggested that a horizontal stabilizer attached to the aft end of the pendulum would aerodynamically offset that shift. The horizontal stabilizer needs to be added to this design.

Up to this point, the discussion of the PNR has shown the nose rudder to be parallel to the physical pendulum. With this arrangement, the PNR is intended to control yaw in the kite to maintain a down-wind position of the kite and the tether. The operation of the MRK system requires a significant departure from this position. The azimuth of the kite and the tether will be required to approach $90°$. The full function of the PNR can be obtained at azimuths up to $90°$ in either direction by altering the angle between the nose rudder and the pendulum, as shown in Figure 6. The angle, θ, can be manually preset before launch, or it can be under radio control during flight. As θ, approaches $90°$, the controlled azimuth of the kite will also approach $90°$. The PNR delta kite is a work in progress. Development will continue in the future.

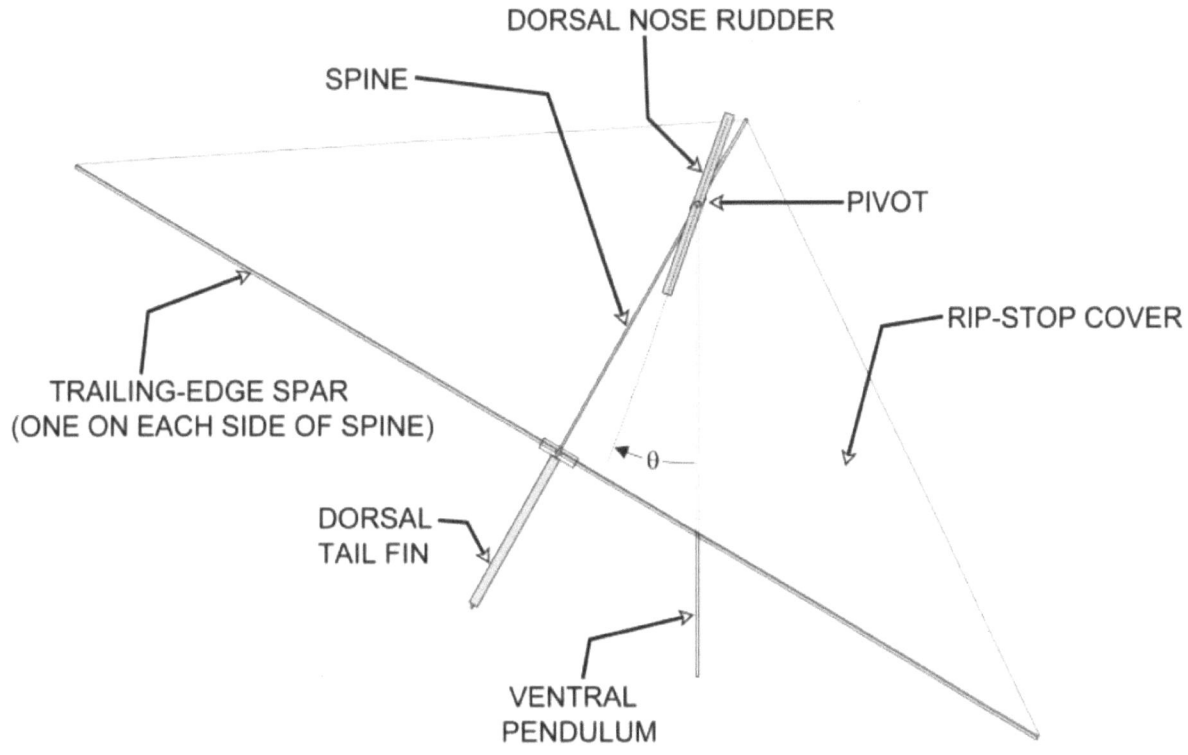

**FIGURE 6. DORSAL SIDE OF PNR DELTA KITE SHOWING
THE ANGLE BETWEEN THE RUDDER AND THE PENDULUM**

In summary, MRK air-vehicle stability is expected to be controlled in pitch by keeping the tether attachment point forward of the centers of lift and gravity, in roll by the use of an adjustable bridle (perhaps with the use of a rate gyro), and in yaw by any one or perhaps all of the three suggested approaches. Firstly, using the feed back of a rate gyro to control a rudder, secondly, by using a sequence of diving and climbing maneuvers (sheeting), and thirdly, by using the PNR concept.

Appendix G. The Dbl And The Horizontal Axis Wind Turbine

We wish to consider the power available from all of the air captured by a static aperture. The density of kinetic energy (energy, E, per unit volume, V) in air flowing at velocity v_W is:

$$\frac{dE}{dV} = \frac{1}{2}\rho_a v_W^2 .$$

This is the energy density available when the air is just brought to a stop.

Now consider a static aperture of area S, used to collect the air. Air is collected by the aperture at the rate Sv_W units of volume per unit of time. If we multiply Sv_W by the energy density (above), we get the power P_{STOP} available by halting the captured air (i.e. by capturing its kinetic energy):

$$P_{STOP} = \frac{1}{2}\rho_a S v_W^3 .$$

Systems anchored to a fixed point on the Earth's surface can capture no more wind power than P_{STOP}. In addition, such systems are limited by an efficiency factor. In the case of the Horizontal Axis Wind Turbine (HAWT), that factor is the Betz limit, 59.3%. Applying the Betz factor to the above equation, the upper limit for the HAWT is

$$P_B = \frac{0.593}{2}\rho_a S v_W^3$$

The density of air at sea level is 1.22 Kg/m^3 and using wind speed of 6 m/s, this becomes

$$P_B = 78.13S$$

in MKS units.

If we let P_B be equal to the power developed by the DBL in 6 m/s wind, then the above equation can be solved in terms of S, the area of a Betz-limited HAWT required to obtain the same power. The DBL was analyzed for power yield using the EVA, and the corresponding HAWT area was calculated. The result is shown in Figure 1 below. At the load limit for the air vehicles, the DBL power generation (about ½ MW at 6 m/s wind speed) is equivalent to that from 5,900 m^2 of HAWT. That is almost 1½ acres. Commercial HAWT systems are rated at much higher wind speeds. So, a commercial HAWT capable of delivering ½ MW at 6 m/s wind speed would be rated at 2½ MW, and the rotor diameter would be 100 m.

If the structural limitations of the DBL (air-vehicle load limit, lateral skidding forces and lift-off) could be overcome, then the maximum DBL power could be realized. That maximum is equivalent to the power from 26,409 m^2 of HAWT. That is in excess of 6½ acres. That equivalent area is the area swept out by the HAWT rotors. Keep in mind that the total wing area from all four air vehicles in the DBL is 67.4 m^2.

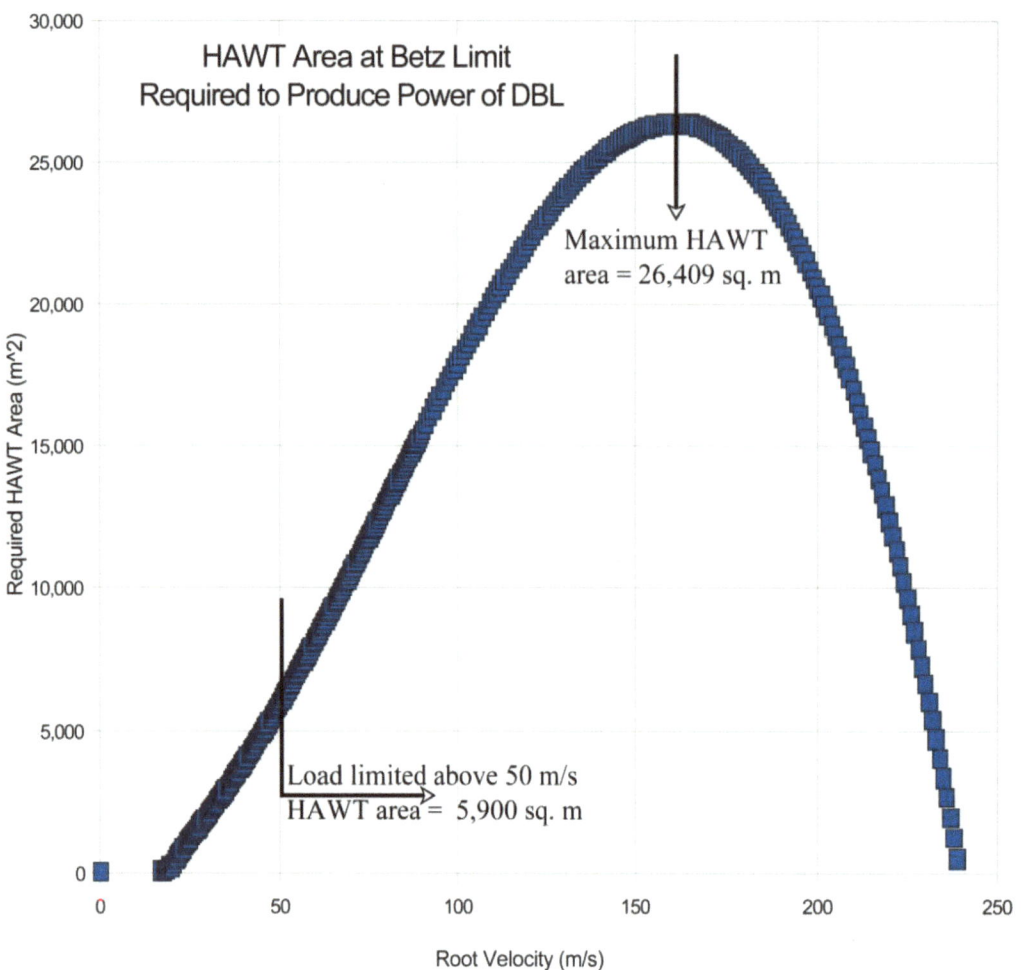

Figure 1. HAWT Sweep Area Required to Produce the Same Power as the DBL

Appendix H Glossary Of Acronyms

BYLB	Bright Young Light Bulb
CFRP	Carbon Fiber Reinforced Plastic
DBL	Design Baseline
DBL02	Second Design Baseline
DME	Dimethyl Ether
DS	Dynamic Soaring
EVA	Equilibrium Vector Analysis
HAWT	Horizontal Axis Wind Turbine
KPH	Kilometers Per Hour
L/D	Lift Drag Ratio
MPH	Miles Per Hour
MRK	Moving Root Kite
PNR	Pendulum Nose Rudder
RC	Radio Control
RMS	Root Mean Squared
TCUAV	Train Cable Unmanned Air Vehicle
UAV	Unmanned Air Vehicle

ABOUT THE AUTHOR

Dennis Wakefield Stevens

Mr. Stevens took a bachelors degree in physics in 1959 and has 36 years industrial experience in materials science and aerospace engineering. He has 17 publications reviewed by peers and four U.S. patents. He taught Advanced Engineering Mathematics at San Diego State University.

www.ingramcontent.com/pod-product-compliance
Lightning Source LLC
Chambersburg PA
CBHW050720180526
45159CB00003B/1085